YOUR KNOWLEDGE HAS VALUE

- We will publish your bachelor's and
  master's thesis, essays and papers

- Your own eBook and book -
  sold worldwide in all relevant shops

- Earn money with each sale

Upload your text at www.GRIN.com
and publish for free

Gerhard Langenberger (Hrsg.)

LILAC Publications

Band 1

Soil classification in the Naban River Watershed National Nature Reserve

GRIN Publishing

Bibliographic information published by the German National Library:

The German National Library lists this publication in the National Bibliography; detailed bibliographic data are available on the Internet at http://dnb.dnb.de .

Imprint:

Copyright © 2010 GRIN Verlag, Open Publishing GmbH
Print and binding: Books on Demand GmbH, Norderstedt Germany
ISBN: 978-3-640-70013-4

This book at GRIN:

http://www.grin.com/en/e-book/157695/soil-classification-in-the-naban-river-watershed-national-nature-reserve

Preface

The group 'Living Landscapes China' provides a platform for people interested in and working on land-use cover change and its implications in SW China, especially Yunnan Province. It follows a holistic approach and covers socio-economic as well as environmental topics. Everybody with substantial experience and data on topics related to land-use change is invited to present his or her data and make it accessible to the general public.

LILAC Publications

'LILAC Publications' intends to safe valuable base-line data acquired during the Sino-German Research Cooperation "Conservation of cultural landscapes through diversification of resource use - strategies and technologies for agro-ecosystems in mountainous Southwest China", in brief, the "Living Landscapes China" or "LILAC" project (www.lilac.uni-hohenheim.de), and make it accessible to a wider audience.

The LILAC project was funded by the German Federal Ministry of Education and Research (BMBF: FK 0330797) and the Chinese Ministry of Science and Technology (MoST: 2007DFA31660).

Background

Southwest China has undergone serious changes in recent years. In this respect, it is representative of large areas in South Asia and in particular the ‚Greater Mekong Subregion' (GMS) with its 300 Mio inhabitants. The GMS is an informal cooperation of Thailand, Lao PDR, Myanmar, Cambodia, Viet Nam, and the P.R. China (Yunnan and Guangxi Province) to promote the development of the region.

While being far off from global developments and characterized by subsistence economy in the past, the region turned out to become one of the most dynamic areas worldwide. A new highway system connects the countries of the GMS, and the area is opened up to international markets. Cash crops such as rubber, which plays already a significant role in local economy, will become even more attractive and speed up land-use cover change (LUCC) with all its consequences for man and his environment.

As a consequence of the dynamic development, socio-economic and cultural traditions are challenged, leading to a rapid change or even disappearance of traditional land-use systems. The consequences are manifold and affect people and their livelihood, as well as the century-old cultural landscape with its high (agro-) biodiversity.

G. Langenberger (Editor)

University of Hohenheim

Institute of Plant Production and Agroecology in the Tropics and Subtropics (380b)

70593 Hohenheim, Germany

Soil classification in the Naban River Watershed National Nature Reserve

Report

Maria Wolff / Zhang Lulu

Content

List of figures

List of tables

1 The Naban River Watershed National Nature Reserve

1.1 General introduction

The Naban River Watershed National Nature Reserve (NNNR) is one of several nature reserves in the Dai Autonomous Prefecture of Xishuangbanna in the south of Yunnan province. The prefecture is located between 21°8`N to 22°36`N and 99°56`E to 101°50`E covering an area of 19.223 km^2. In the southwest and south east Xishuangbanna borders Myanmar and Laos (Fig.1).

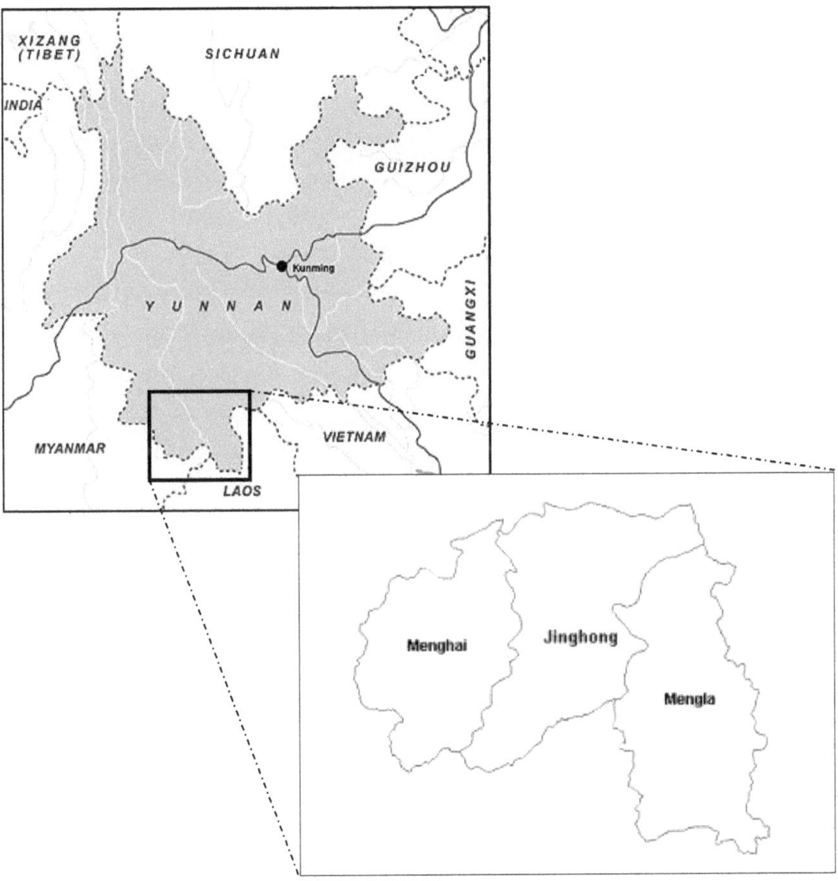

Figure 1: Location of Xishuangbanna and the counties Menghai, Jinghong, Mengla

Yunnan is shaped by the southern part of the Hengduan Mountains (APEL, 1996; YUNNAN SOCIETY OF ECOLOGICAL ECONOMICS et al., 1992). Approximately 95 % of the province consists of mountains and hills, only 5 % are flat plains and river basins (WANG, 2000, with elevations ranging between 420 and 2,400 m asl.

In Xishuangbanna typical land formations are mountainous and hilly areas combined with plains. Large valleys can also be found here, which are mainly surrounded by hilly areas or single hills (ACHILLES, 1997; YUNNAN SOCIETY OF ECOLOGICAL ECONOMICS et al., 1992).

The major river of the area is the Mekong River or Lancang Jiang with its headwaters in Tibet. The Mekong catchment is characterized by several tributaries, such as the rivers Luoso, Nanla and Liusha. In total, 2,762 rivers are influencing the variety of landscapes in Xishuangbanna (WU & OU, 1995; YUNNAN SOCIETY OF ECOLOGICAL ECONOMICS et al., 1992; ZHANG, 1986).

The nature reserve covers an area of 267 km² and is located 20 km northwest of Jinghong City, the prefecture capital of Xishuangbanna. It extends between 22°04`N to 22°17`N and 100°32` to 100°44`E. The reserve encompasses an altitude range of 539 m to 2,304 m a.s.l., with an estimated 90% of the area situated between 600 m and 1500 m a.s.l. The western and central parts of the reserve belong to the Naban River watershed, whereas the eastern slopes drain directly into the Mekong River.

1.2 Geology

Geologically, the reserve can be divided into two parts. The western part is dominated by granite as the main parent material, while the eastern part mainly shows phyllites.

1.3 Soils and nutrients

The investigated area is located in the so called Red Soil Region of China. With an area of 2.6 Mio km^2 this region accounts for nearly 20 % of China`s total area (WILSON et al. 2004) and is mainly distributed in the subtropics and tropics of China.

Chinese Red Soils build up four groups: Latosols, Lateritic red earths, Red earths and Yellow earths. Table 1 gives an overview of the groups and their equivalences according to the World Reference Base for Soil Resources (WRB) (FAO/UNESCO 1998).

Table 1: Red soils in China (Source: FAO/UNESCO 1998[1]; SHI et al. 2006[2]; HE et al. 2004a[2])

Chinese soil classification 1984		WRB 1998[1]	US Soil Taxonomy 1994[2]	US Soil Taxonomy 1999[3]
Order	Main Group	Reference Soil Group	Order	Suborder
Red Soils	Latosols	Plinthosols, Ferralsols, Acrisols, Alisols, Nitisols, Lixisols, Cambisols	Oxisolos, Ultisols	Udox. Udult, Humult
	Lateritic Red Earths		Inceptisols, Oxisols, Ultisols	Udult, Ochrult Ochrept
	Red Earths		Alfisols, Inceptisols, Ultisols, Vertisols	Udult, Udalf, Ochrept
	Yellow Earths		Inceptisols, Ultisols	Udalf, Umbrept, Ochrept, Udent

Chinese red soils are subtropical and tropical soils which are strongly weathered, thus showing strong leaching of nutrients. The soils are characterised by high contents of so-called low active clays as well as accumulation of Al and Fe oxides and hydroxides. A commonly low content of organic substances (OS) can lead to a lack of total nitrogen (TN). Red soils often show a deficiency of plant available P, due to the high content of Al and Fe oxides as well as kaolinite causing a strong adsorption of P (He et al. 2004b). As a result of the already mentioned strong weathering and leaching the soils have low cation exchange capacity (CEC) and low base saturation (BS) as well as low water-holding capacity (WHC) (HE et al. 2004a, WILSON et al. 2004).

Table 2: Nutrients distribution of red soils in China (Source: HE et al. 2004a)

Soil Property		Latosols	Lateritic red soils	Red earths	Yellow earths
Clay	%	40-60	30-40	25-40	15-30
pH		4.6-5.4	5.0-5.5	4.2-5.9	4.5-5.5
CEC	cmol$_c$/kg	2.8-12.6	5.4-11.7	5.3-8.9	10.0-20.0
BS	%	15-20	20-30	15-25	15-60
Total N	g/kg	0.5-1.2	0.8-1.5	1.4-2.0	1.8-3.3
Total P	g/kg	0.3-1.5	0.5-0.9	0.6-0.9	0.5-1.9
Total K	g/kg	0.5-6.4	15-25	15-20	11-26
Hydro K	mg/kg	25-170	-	60-170	-
Avail. P	mg/kg	2-10	2-9	2-4	-
Avail. K	mg/kg	31-65	50-130	37-120	-
Profile		A B$_s$ B$_v$ C	A B C	A B$_s$ B$_v$ C	A (AB) B C

According to ANONYMUS (1987, in APEL, 1996) five major soil types were found in Xishuangbanna, which can be classified as three zonal (dependent on altitude) and two azonal types. The zonal soil types are „brick-red soils" at altitudes of 600 to 900 m a.s.l., „yellow-red soils" or „red soils" between 900 to 1.600 m a.s.l., and „red mountain soil" above 1.600 m a.s.l.. Applying the FAO-Classification all three soil types are Ferralsols with different contents of iron oxides (APEL, 1996).

APEL (1996) states that azonal soil types are calcaric soils and purple soils. Purple soils are according to WRB Regolsols derived from devonian clay shale and show high contents of mangan. The non zonal soils show only very small distributions in Xishuangbanna. Next to the named soil types also alluvial soils as well as Gleysols and Planosols can be found in valleys and depressions (BRUENIG et al., 1986).

2 Material and methods

2.1 Field investigations

Field investigation and soil sampling were conducted in November 2009.

2.1.1 Site selection

The aim of the investigation was to characterize typical soil types in dependence of the geomorphology (altitudes, geology) and selected land use types (forest, rubber plantation, paddy field, waste land) in the NRWNNR. As forests are a target land use type investigated in the LILAC project the influence of geomorphology was mainly assessed under forests of different *altitudes* by setting up a *transect 1* in the western part of NRWNNR from Hui Lao Xin Zhai (approx. 1000 m a.s.l.) to Beng Gang Ha Ni (approx. 1700 m a.s.l.). Whereas the influence of *land use* types were investigated in the eastern part of NRWNNR by analysing a *transect 2* around the villages Na Ban and Man Dian. Here all profiles were located at altitudes between 660 to 703 m a.s.l..

2.1.2 Soil profile description

After establishing the soil profiles all characteristics were noted and the profiles were described by following the WRB guidelines of soil description (FAO, 2006a) as well as the German soil description guidelines KA5 (Ad hoc AG Boden, 2005).

2.1.3 Soil sampling

After determining the horizons at each profile three mineral soil samples were taken at each horizon by using three soil corers with a volume of 100 cm^3. The three samples were then given into a plastic bag and thoroughly homogenized to a mixed sample. After weighing all fresh samples, the plastic bags were opened and stored at the research station in Naban village for air drying before transportation to the laboratory for following analysis. Here the samples were then oven-dried for 48 h at 40 °C.

2.2 Laboratory analysis

2.2.1 Bulk density, fine earth density, water content

The bulk density (BD) was determined by weighing the field-wet samples and calculated by using the known volume of the soil corers:

$$BD = \frac{fresh\ weight\ (g)}{volume\ (cm^3)} \tag{1}$$

After drying the samples were passed through a 2 mm sieve to remove roots and gravels >2 mm. The fine material was weighed in order to calculate the fine earth density (BD_f) with:

$$BD_f = \frac{air\ dried\ weight\ (g)}{volume\ (cm^3)} \tag{2}$$

The soil water content (WC) is is the ratio of the mass of water in the soil samples to the mass of dried soil. It is determined by drying method by 105 °C and calculated as follows:

$$WC\ (\%) = \frac{M_w - M_d}{M_w}\ x\ 100 \tag{3}$$

with   WC = Gravimetric water content (%)

M_w = Mass of the air-dried soil sample (g)

M_d = Mass of the soil samples dried with 105 °C (g)

2.2.2 Texture

By using the combined sieving and pipetting method (DIN ISO 11277) the particle size fractions of the soil samples were determined. Therefore, organic matter was dissolved with 30 % H_2O_2 with a following dispersion with $Na_4P_2O_7$ solution. Dissolving of the $CaCO_3$ contents were not applied because of the insignificant quantities of carbonate (< 2 %).

2.2.3 pH value

The pH value was measured with a glass electrode. The soil was thoroughly mixed with deionized water with a ratio of 1:2.5.

2.2.4 Soil organic carbon (SOC) and total nitrogen (TN)

For determining the contents of total carbon (TC) and total nitrogen (TN) the soil samples were dried at 40 °C and milled before applying a complete dry combustion with a CNS analyzer (Vario EL III / elementar, Heraeus). According to the measured ph values it was assumed that the content of carbonates could be neglected. To verify this assumption the carbonate contents of randomly chosen samples showing pH values above 5 were measured. Here all contents were below 2 %, confirming our presumption. Thus, the total carbon content (TC) corresponds to the content of soil organic carbon (SOC).

2.2.5 Nutrients

Aliquots of all soil samples were dried at 105°C. Subsequently HNO_3 (65 %), HF (40 %) and $HClO_4$ (70 %) were added for using the microwave decomposition before determination of element contents at an ICP AES (Spectro Ciros).

2.2.6 Cation exchange capacity

The effective cation exchange capacity (ECEC) has been measured by exchanging the cations with 0.5 M NH_4Cl solution and subsequent filtration through a 0.45 µm membrane filter. The extract has been analysed with the ICP AES Spectro Ciros equipment to determine the ECEC as the sum of the hydrogen (H+) and the positively charged elements (K^+, Na^+, Ca^{2+}, Mg^{2+}, Mn^{2+}, Al^{3+} and Fe^{3+}) per g soil. The base saturation (BS) is the sum of the percentages of the base cations (K^+, Na^+, Ca^{2+} and Mg^{2+}) (SCHEFFER and SCHACHTSCHABEL, 1998).

3 Results

3.1 World Reference Base

The World Reference Base for Soil Resources (WRB) is the successor to the International Reference Base for Soil Classification (IRB). Its task is to apply the IRB principles of definition and linkages to the existing classes of the Revised FAO-Unesco Soil Map of the World Legend (FAO/UNESCO, 1988) (SPAARGAREN and DECKERS, 1998).

The main objective of the World Reference Base for Soil Resources is to provide scientific depth and background to the Revised Legend, so that it incorporates the latest knowledge relating to global soil resources and interrelationship. Specially the following objectives (SPAARGAREN and DECKERS, 1998):

- to develop an internationally acceptable framework for soil resources, to which national classification systems can be related and through which the national systems can be linked;
- to enable the international use of pedological data, not only by soil scientists, but also by other users, such as geologists, botanists, agronomists, ecologists, foresters, farmers, etc.;
- to acknowledge important lateral relationship of soils and soil horizons as characterized by topo- and chronosequences; and
- to emphasize the morphological characterization of soils rather than to follow an approach purely based on laboratory analyses.

In the WRB soil characteristics, properties and layers, which are in combination, are used to describe and define the reference soil groups and soil units. Soil characteristics are parameters, observed or measured either in the field or the laboratory, including color, texture of soils, features of biological activity, voids, mottles as well as analytical measuring (pH, particle distribution, exchangeable cations, exchangeable cation capacity, ...) (SPAARGAREN and DECKERS, 1998).

Soil properties are combinations of soil characteristics and considered to be indicative of soil forming processes (e.g. vertic properties are a combination of heavy texture, smectitic mineralogy, gilgai, slickenside, hard consistence when dry, sticky consistence when wet, shrinking when dry and swelling when wet) (SPAARGAREN and DECKERS, 1998).

Soil horizons are certain depths, which are characterized by one or several properties with a certain degree of expression. The thickness of each horizon is varied, ranging from a few centimeters to several meters, and the boundaries of horizons are diffuse, gradual, clear or abrupt (SPAARGAREN and DECKERS, 1998).

Reference soil groups (RSG) are defined by a vertical combination of horizons within a certain depth. In the WRB 30 reference soil groups were firstly proposed in 1998, namely Histosols, Cryosols, Anthrosols, Leptosols, Vertisols, Fluvisols, Solonchaks, Gleysols, Andosols, Podzols, Plinthosols, Ferralsols, Planosols, Solonetz, Chernozems, Kastanozems, Phaeozems, Gypsisols, Durisols, Calcisols, Albeluvisols, Alisols, Nitisols, Acrisols, Lixisols, Umbrisols, Cambisols, Arenosols and Regosols (SPAARGAREN and DECKERS, 1998).

3.2 Soil profile description

3.2.1 Profile 1

P1: Hui Lao Xin Zhai

Ferralic Cambisol (sodic, humic, eutric, chromic)

According to WRB (FAO, 2006) a *cambic* horizon was defined as diagnostic horizon. As prefix *ferralic* was chosen because of a very low ECEC. *Sodic* and *eutric* were identified as suffixes indicating a higher percentage of Na and Mg on the exchange complex and a base saturation > 50 %.

Figure 2: Profile P1

Table 3: Soil properties of P1

Elevation asl	1100 m
Position	Middle slope
Slope gradient*	10 (> 60%)
Land use	Forest
Parent material	Granite

* FAO/UNESCO 1998

Horizon	Depth (cm)	Soil Colour
Oi	2	
Ah	0-8	5YR 3/4
Ah-Bw1	8-26	7.5 YR 4/4
Bw1	26-48	7.5 YR 5/8
Bw2	48-78	5 YR 5/6
Bw2-C	78-	5 YR 5/8

Horizon	Coarse Fraction (%)	BD (g/cm^3)	Texture	Clay (%)	pH (H$_2$O)	TC (%)	ECEC cmol$_c$/kg	BS (%)
Ah		1.25	Sandy clay loam	28.34	6.29	3.46	15.20	99.53
Ah-Bw1		1.43	Clay loam	32.44	5.60	1.95	7.17	92.98
Bw1	< 2	1.54	Sandy clay loam	28.99	5.28	1.17	4.95	59.25
Bw2	10	1.51	Sandy clay loam	27.42	5.27	0.80	3.82	50.37
Bw2-C	30	1.49	Sandy clay loam	24.91	5.37	0.58	3.66	49.33

3.2.2 Profile 2

P2: Gei Yang Gong Di

Vetic Ferralsol (alumic, hyperdystric, clayic, xanthic)

Here a *Vetic Ferralsol* was classified, as this profile shows a ECEC of less than 6 cmolc/kg clay. The effective Al saturation is more than 50 %, consequently the suffix *alumic* was used. Reflecting the very low base saturation the suffix *hyperdistric* was applied.

Figure 3: Profile P2

Table 4: Soil properties of P2

Elevation asl	1664 m
Position	Upper slope
Slope gradient*	6 (5-10%)
Land use	Waste land
Parent material	Granite

* FAO/UNESCO 1998

Horizon	Depth (cm)	Soil Colour
Oi	1	
Ah	0-15	10YR 4/6
Bws-Ah	15-25	7.5 YR 5/8
Bws1	25-43	7.5 YR 6/8
Bws2	43-70	5 YR 6/8
Bws2-C	70-	5 YR 6/8

Horizon	Coarse Fraction (%)	BD (g/cm^3)	Texture	Clay (%)	pH (H$_2$O)	TC (%)	ECEC cmol$_c$/kg	BS (%)
Ah	< 2	1.13	Clay loam	39.26	4.81	4.28	5.11	21.57
Bws1-Ah	5	1.43	Clay	42.44	4.83	2.11	2.78	11.26
Bws1	5	1.61	Clay	48.32	4.90	1.34	1.92	6.52
Bws2	5	1.60	Clay	46.16	5.14	0.86	1.20	11.51
Bw2-C	5	1.56	Clay loam	35.67	5.26	0.41	0.93	11.99

3.2.3 Profile 3

P3: Beng Gang La Hu

Haplic Lixisol (humic, clayic, chromic)

Figure 4: Profile P3

Showing a clear increase in clay content an *argic* horizon was identified. But reflecting the high base saturation an *Acrisol* could not been classified. Thus, the soil type *Lixisol* had to be chosen with the suffixes *humic, clayic* and chromic, indicating organic carbon contents of > 1 %, a texture of clay and a soil colour (Munsell hue) redder than 7.5 YR.

Table 5: Soil properties of P3

Elevation asl	1560
Position	Middle Slope
Slope gradient*	10 (> 60%)
Land use	Forest
Parent material	Granite

* FAO/UNESCO 1998

Horizon	Depth (cm)	Soil Colour
Oi	7	
Root felt	4	
Ah	0-15	7.5 YR 4/2
Ah-Bt1	15-23	5 YR 4/4
Bt1	23-50	10 YR 4/6
Bt2	50-	5 YR 4/6

Horizon	Coarse Fraction (%)	BD (g/cm^3)	Texture	Clay (%)	pH (H$_2$O)	TC (%)	ECEC cmol$_c$/kg	BS (%)
Ah		1.56	Clay loam	34.22	4.75	2.06	5.60	53.09
Ah-Bt1		1.56	Clay loam	36.84	4.89	3.21	5.99	58.26
Bt1		1.43	Clay	46.39	5.03	1.51	5.84	49.11
Bt2	< 2	1.56	Clay	42.64	5.17	1.19	5.02	49.43

3.2.4 Profile 4

P4: Beng Gang Ha Ni

Ferralic Cambisol (alumic, dystric, epiclayic, chromic)

Figure 5: Profile P4

A *Ferralic Cambisol* was classified for P 4. The suffixes. alumic and *dystric* were identified indicating a higher percentage of Al on the exchange complex and a base saturation < 50 %.

Table 6: Soil properties of P4

Elevation asl	1730
Position	Upper Slope
Slope gradient*	6-7 (5-15%)
Land use	Forest
Parent material	Granite

* FAO/UNESCO 1998

Horizon	Depth (cm)	Soil Colour
Oi	6	
Worm activities	2	7.5 YR 4/3
Ah	0-14	10 YR 4/6
Bw1	14-46	10 YR 4/6
2Bw2	46-96	7.5 YR 5/8
2Bw2-C	96-	10 YR 5/6

Horizon	Coarse Fraction (%)	BD (g/cm^3)	Texture	Clay (%)	pH (H$_2$O)	TC (%)	ECEC cmol$_c$/kg	BS (%)
Ah		1.37	Clay	47.68	4.06	3.15	6.74	6.48
Bw1		1,42	Clay	42.97	4.39	1.64	4.73	3.82
2Bw2		1,54	Sandy clay loam	30.89	4.64	0.64	4.93	8.62
2Bw2-C		1,44	Sandy loam	15.79	4.79	0.27	3.85	2.92

3.2.5   Profile 5

P5: Cha Chang

Ferralic Cambisol (alumic, dystric, siltic, chromic)

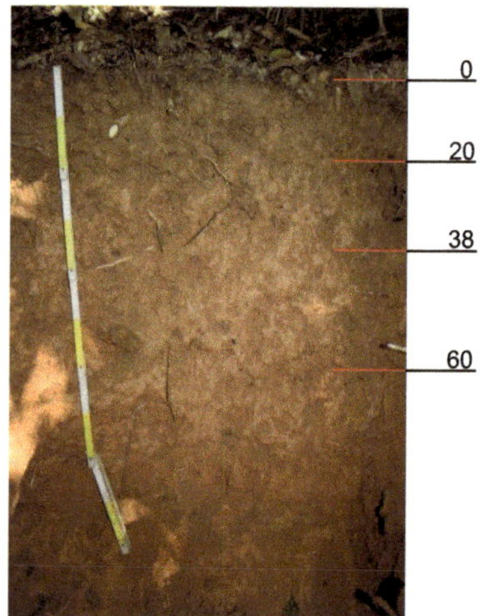

This profile was also classified as a *Ferralic Cambisol*, showing a low ECEC and base saturation (*dystric*) with a high Al saturation (*alumic*).

Figure 6: Profile P5

Table 7: Soil properties of P5

Elevation asl	689 m
Position	Upper slope
Slope gradient*	10 (> 60%)
Land use	Forest
Parent material	Phyllite?

* FAO/UNESCO 1998

Horizon	Depth (cm)	Soil Colour
Oi	6	
Root felt	3	
Ah-Bw1	0-20	10 YR 5/6
Bw1-Ah	20-38	7.5 YR 5/8
Bw1	38-60	7.5 YR 5/8
Bw2	60-	7.5 YR 5/8

Horizon	Coarse Fraction (%)	BD (g/cm^3)	Texture	Clay (%)	pH (H$_2$O)	TC (%)	ECEC cmol$_c$/kg	BS (%)
Ah-Bw1		1.20	Clay	39,64	4.05	1.94	6.18	8.56
Bw1-Ah		1.44	Clay	39.83	4.39	1.25	4.86	4.36
Bw1		1.59	Silty clay	40.29	4.40	0.75	4.66	3.82
Bw2		1.66	Silty clay	43.40	4.41	0.55	4.68	3.74

3.2.6 Profile 6

P6: Na Ban, rubber old

Vetic Plinthic Acrisol (alumic, humic, profondic, clayic, chromic)

Figure 7: Profile P6

With a clear *argic* horizon with a base saturation < 50 % this profile could be classified as an *Acrisol.* In the depth between 60 and 77 cm a *plinthic* horizon was found that consists of cemented nodules and irregular aggregates in which Fe is an important cement (FAO, 2006b). The prefix *vetic* shows a low ECEC per clay, whereas the suffix *alumic* states the high Al saturation.

Table 8: Soil properties of P6

Elevation asl	683 m
Position	Middle slope
Slope gradient*	9 (30-60%)
Land use	Rubber trees
Parent material	Phyllite?

* FAO/UNESCO 1998

Horizon	Depth (cm)	Soil Colour
Oi	1	
Ah-E	0-10	7.5 YR 4/6
E	10-20	7.5 YR 4/6
Bt1	20-60	7.5 YR 5/8
Bt1c	60-77	5 YR 5/8
Bt2	77-	5 YR 5/8

Horizon	Coarse Fraction (%)	BD (g/cm^3)	Texture	Clay (%)	pH (H$_2$O)	TC (%)	ECEC cmol$_c$/kg	BS (%)
Ah-E		1.56	Silty clay	43.13	4.46	1.81	5.36	18.97
E		1.61	Silty clay	44.28	4.54	1.30	4.88	20.68
Bt1		1.60	Clay	57.00	4.69	1.09	4.94	21.28
Bt1c	20	1.62	Clay	55.91	4.87	0.80	4.24	24.69
Bt2		1.50	H clay	66.76	4.88	0.77	3.78	24.19

3.2.7 Profile 7

P7: Na Ban, paddy

Hydragric Stagnic Anthrosol (sodic, eutric)

Figure 8: Profile P7

The paddy rice soil is a soil with long and intensive agricultural use. The *anthraquic* and *hydragric* horizon have a combined thickness of more than 50 cm. Thus, the soil type could be identified as an *Anthrosol*, showing *stagnic* properties, reflected by intensive mottling.

Table 9: Soil properties of P7

Elevation asl	659 m
Position	terrace
Slope gradient*	1 (0%)
Land use	Paddy rice
Parent material	

Horizon	Depth (cm)	Soil Colour
Apg	0-10	10 YR 3-4/4
Bg	10-55	10 YR 4/4
Btg	55-	10 YR 4/6

* FAO/UNESCO 1998

Horizon	Coarse fract.	BD (g/cm^3)	Texture	Clay (%)	pH (H$_2$O)	TC (%)	ECEC cmol$_c$/kg	BS (%)
Apg		1.65	Clay loam	29.00	5.18	1,24	5,45	91,39
Bg		1.78	Loam	24.63	6.05	1,08	7,94	99,44
Btg		1.90	Clay loam	31.46	6.27	1,15	9,26	99,41

3.2.8 Profile 8

P8: Man Dian

Ferralic Cambisol (sodic, eutric, clayic, chromic)

Figure 9: Profile P8

The *Ferralic cambisol* in Profile 8 also shows an increase of the clay content, although here the diagnostic horizon is a *cambic* horizon with a thickness of more than 30 cm. In this case the soil type is specified as *sodic* and *eutric* showing a high percentage of Na and Mg on the ECEC and a high base saturation throughout the profile.

Table 10: Soil properties of P8

Elevation asl	686 m
Position	Middle slope
Slope gradient*	9 (30-60%)
Land use	Forest
Parent material	

Horizon	Depth (cm)	Soil Colour
Oi	2	
Ah	0-4	7.5 YR 4/5
Bw1	4-50	7.5 YR 5/6
Bw2c	50-66	7.5 YR 5/8
Bt-C	66-	7.5 YR 5/8

* FAO/UNESCO 1998

Horizon	Coarse fract.	BD (g/cm^3)	Texture	Clay (%)	pH (H$_2$O)	TC (%)	ECEC cmol$_c$/kg	BS (%)
Ah		1.60	Clay loam	35.73	5.79	1.30	6.18	97.87
Bw1	<1	1.42	Clay loam	33.27	4.99	0.84	3.40	72.39
Bw2c	8	1.42	Clay loam	38.57	4.78	0.65	3.28	46.79
Bt-C	45	1.33	Clay	50.04	4.86	0.78	3.24	63.81

3.2.9 Profile 9

P9: Jiang Bian Zhan

Haplic Acrisol (alumic, hyperdystric, profondic, clayic, chromic)

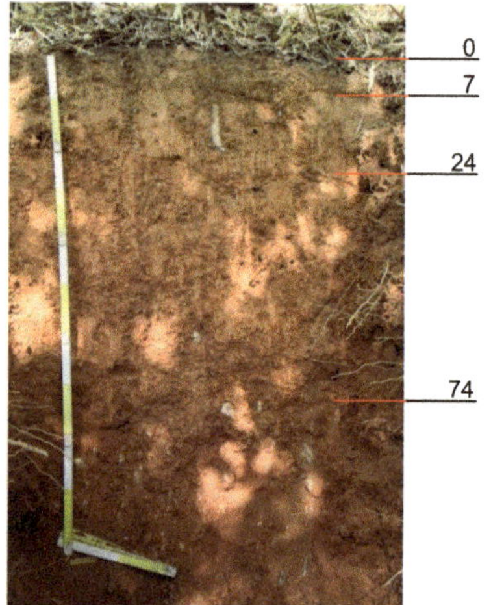

Figure 10: Profile P9

Profile 9 shows an increase in clay content. Consequently an *argic* horizon was selected as diagnostic horizon. Because of the very low ECEC and base saturation an Acrisol was classified according to WRB (FAO, 2006).

Table 11: Soil properties of P9

Elevation asl	686 m
Position	Middle slope
Slope gradient*	9 (30-60%)
Land use	Bamboo forest
Parent material	

* FAO/UNESCO 1998

Horizon	Depth (cm)	Soil Colour
Oi	2	
Ah	0-7	5 YR 5/6
E-Ah	7-24	7.5 YR 5/8
Bt1	24-74	5 YR 6/8
Bt2-C	74-	2.5 YR 5/8

Horizon	Coarse fract.	BD (g/cm^3)	Texture	Clay (%)	pH (H$_2$O)	TC (%)	ECEC cmol$_c$/kg	BS (%)
Ah		1.16	Clay loam	38.43	4.36	3.17	5.57	10.77
E-Ah		1.41	Clay loam	33.17	4.43	1.86	4.87	5.80
Bt1	3	1.41	Clay loam	39.57	4.53	0.82	5.41	5.22
C-Bt2	45	1.39	Clay	46.51	4.71	0.49	6.18	2.16

3.3 Transect 1

3.3.1 Depth and fine earth density

Topsoil

Topsoils under evergreen broadleaved forests reached a thickness of 15 cm. Fine earth densities varied between 0.90 g/cm^3 (profile 2, Gei Yang Gong Di) and 1.16 g/cm^3 (profile 3, Beng Gang Lahu), showing no clear differences between the investigated profiles.

Subsoil

With increasing depth the fine earth densities increased in profiles P1, P2, and P4. Only for P3 no clear trend was visible. With densities between 1.1 g/cm^3 and 1.3 g/cm^3 no obvious differences between the profiles could be detected.

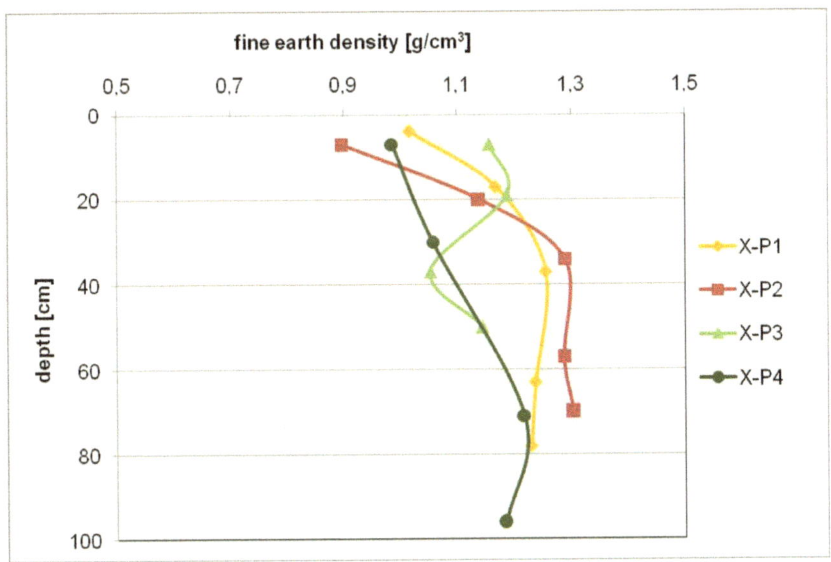

Figure 11: Fine earth densities of all horizons of transect 1

3.3.2 Texture

Topsoil

For all profiles of transect 1 sand (mean: 41.31 %) and clay (mean: 37.39 %) made up the most part of soil particles of Ah horizons. In comparison to all other profiles of this transect the Ah horizon of P4 (Beng Gang Hani) shows a higher clay content with a percentage of 47.68 % (Fig.12). Consequently the topsoils of the profiles were classified with a texture of sandy clay loam and clay loam, whereas P4 shows clay as the dominating textural class.

Subsoil

As Fig. 13 shows the texture of all profiles were dominated by clay and sand contents. In detail Profile 1 had slightly lower contents of clay in the subsoil horizons. Here the main textural class was sandy clay loam. In profiles P2 and 3 the textural class clay was determined, whereas in P4 the classes clay (Bw1), sandy clay loam (2Bw2) and sandy loam (2Bw2-C) where found.

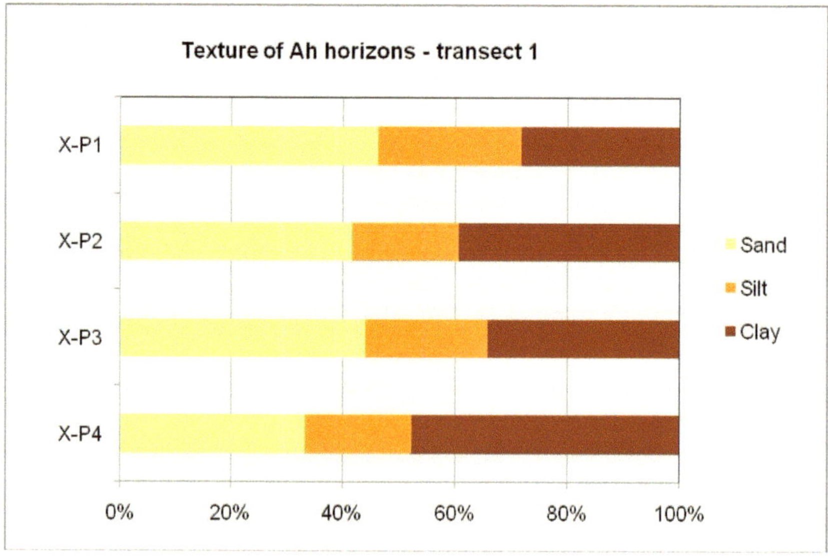

Figure 12: Distribution of soil particles in Ah horizons of transect 1

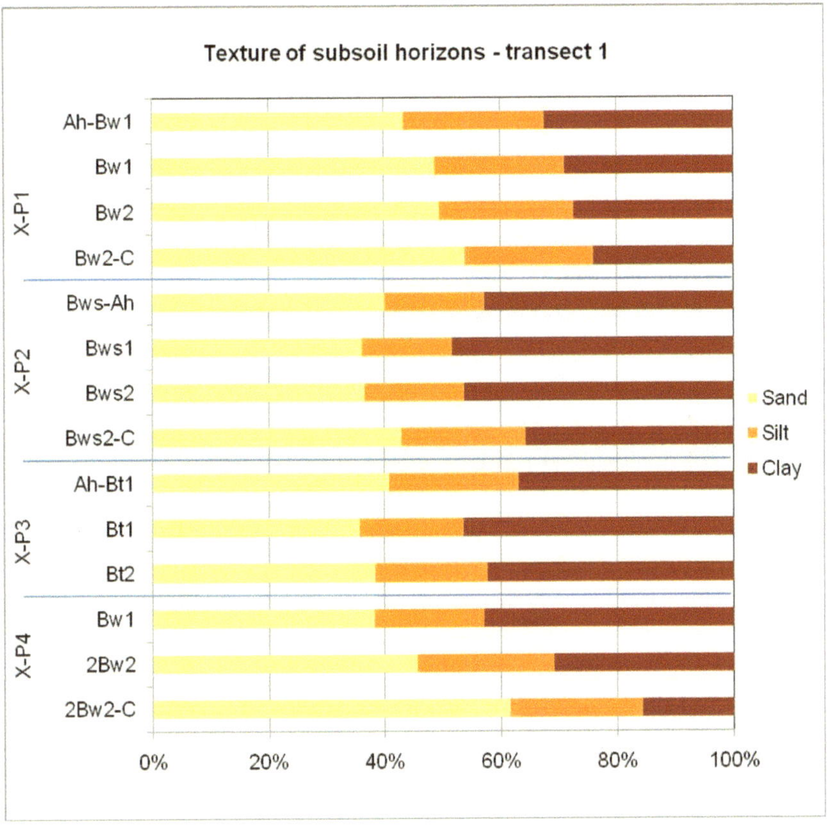

Figure 13: Distribution of soil particles in subsoil horizons of transect 1

3.3.3 pH value

Topsoil

The profiles P2-P4 were showing very acid soil reactions indicated by the low ph in the topsoil horizons (between 4.0 and 5.0). Only for the Ah horizon of P1 a pH value above 6 could be measured, indicating a slightly better living condition for soil microbes. This result also corresponds with the composition of nutrients and exchangeable cations in the Ah horizon of P1 (see 3.3.5, 3.3.6).

Profile P4 had the lowest pH value (4.1), showing clear differences compared to all other profiles of this transect (Fig.14).

Subsoil

In general, the subsoil of P4 was a little more acid than in the other profiles, whereas the pH-values of P4 were again the highest but reached similar values with increased depth. Between P2 and P3 no clear differences could be observed. In addition, the pH-values in the subsoil of P1, P2, and P3 increased with the depth from upper to lower. Only in profile P4 a contradictory trend was observable (Fig.14).

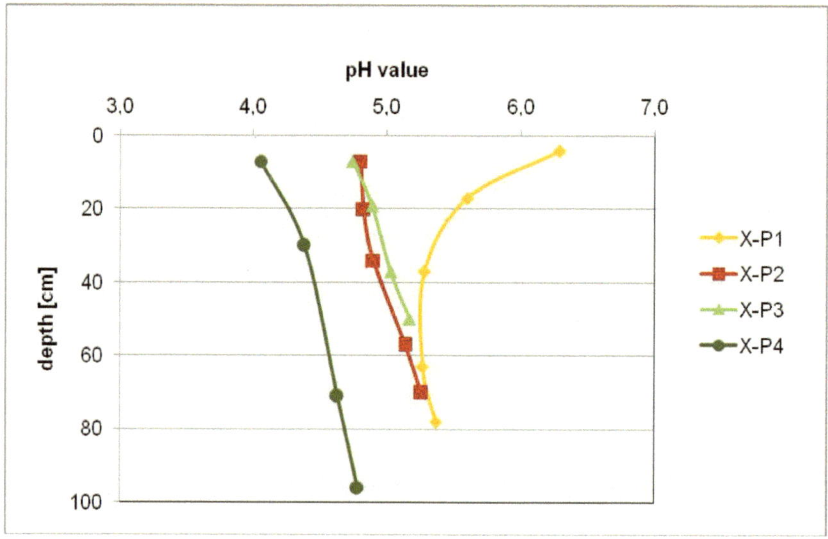

Figure 14: pH value of all horizons of transect 1

3.3.4 SOC and TN

Topsoil

The contents of soil organic carbon (SOC) ranged from 20.6 g/kg (2 %, P3) to 42.8 g/kg (P2) (Fig. 15a). That the highest SOC content was found at the Ah horizon of P2 was unexpected as the site where P2 is located is a cleared wasteland next to the road, whereas all other profiles were established under forest cover. A clear influence of altitude could not be detected.

On the other hand the content of total nitrogen (TN) did not vary in the topsoil horizons between the profiles of transect 1.

The C/N ratio in the topsoil ranged from 9.7 (P3) to 17.9 (P2), indicating slightly unfavourable conditions in the Profile P2 compared to the forest sites.

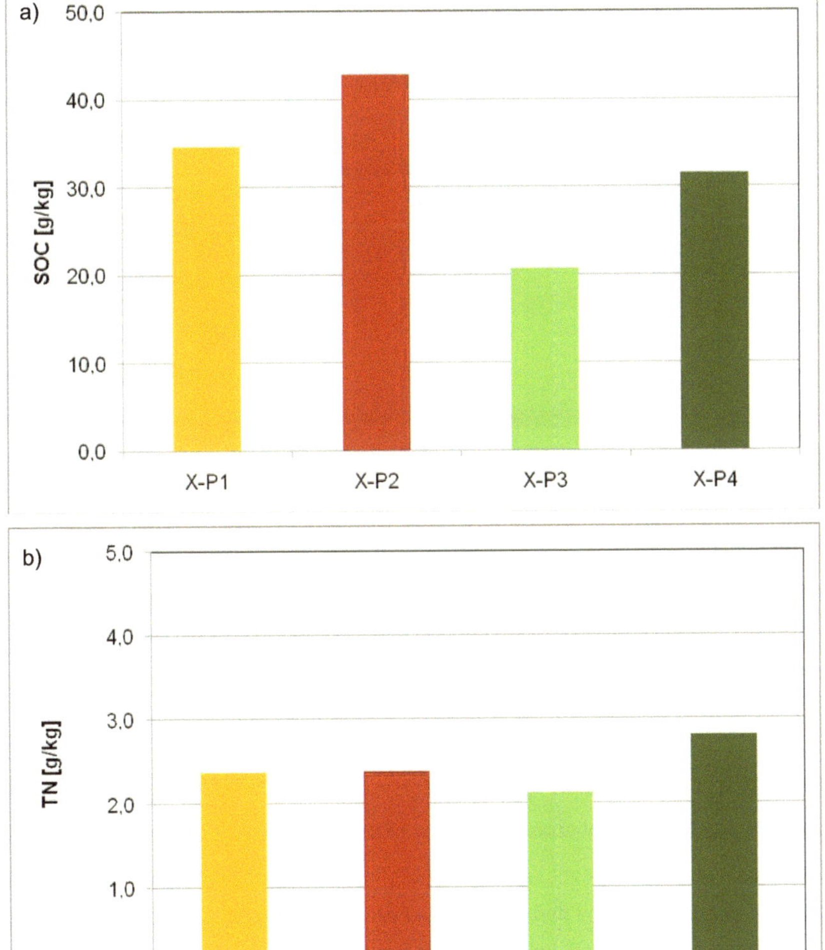

Figure 15: Contents of a) SOC and b) TN of Ah horizons of transect 1

Subsoil

As Clear differences in SOC contents between the profiles of transect 1 were only visible in topsoil horizons and the first (mostly) transitional horizons (see Fig 15a and 16), the following subsoil horizon showed all similar contents and trends of decreasing SOC contents with increasing depth. Especially for the first transitional horizon (Ah-Bt1) but also still visible at the underlying horizons P3 showed the highest SOC contents, indicating a relocation of the SOC from topsoil into deeper horizons at this site.

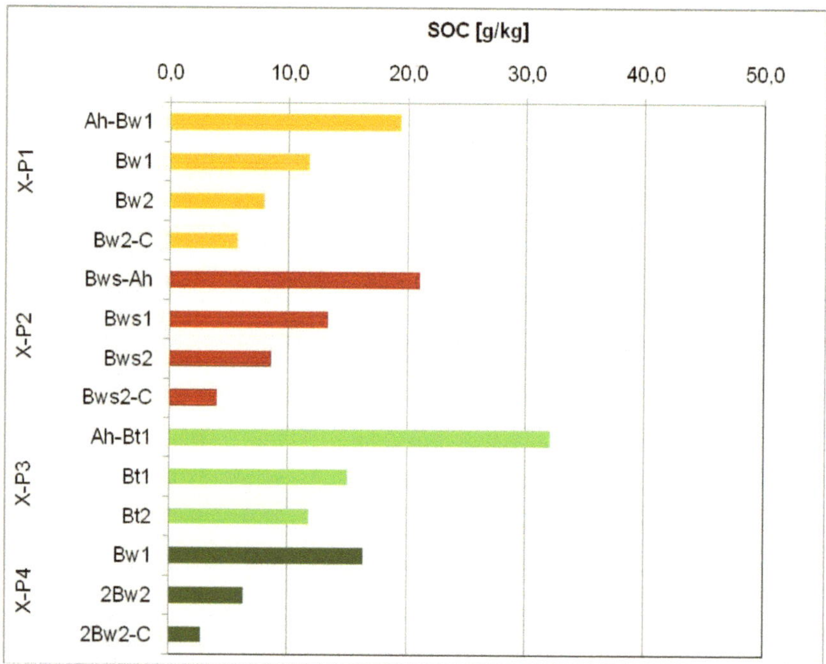

Figure 16: SOC content of subsoil horizons of transect 1

The TN contents of subsoil horizons showed similar trends as the SOC contents with the highest contents found for P3 and decreasing contents with increasing depth (Fig. 17). Again the most obvious differences in the TN contents were found in the transitional horizons from topsoil to subsoil. As already seen for SOC here also P3 had the highest contents with 2.9 g/kg at the Ah-Bt1 horizon.

The SOC/TN ratios in the subsoil horizons did not change significantly compared to the topsoil horizons, thus showing the same trend as was found for the topsoil.

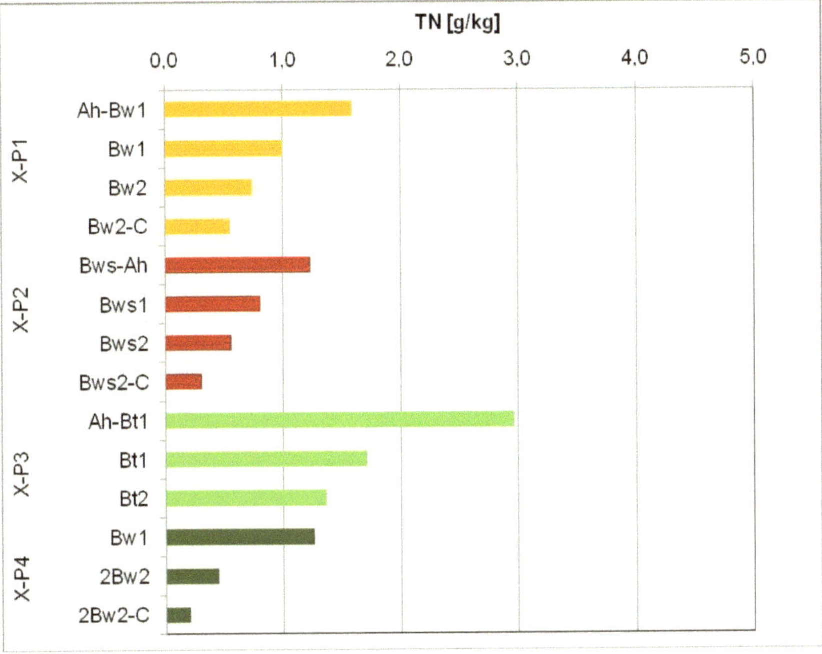

Figure 17: TN content of subsoil horizons of transect 1

3.3.5 Nutrients

Topsoil

The total contents of nutrients are shown in Tab. 12. Comparing only the Ah horizons P1 shows the highest total contents of the elements P, K, Ca, Mg, and Mn, with clear differences especially in total contents of calcium and magnesium in comparison to the other profiles.

Subsoil

Looking on the whole profiles it seems that P2 shows a slight accumulation of iron, whereas the P, K, Ca, Mg, Na, and Mn clearly had the lowest contents in this profile. This might indicate a stronger leaching of nutrients at this site, reflecting the missing forest cover compared to the other sites and profiles.

Comparing P1 with profiles P3 and P4 the differences in total contents of nutrients seems less then found for the topsoil horizons in these profiles. This result may lead to the theory that especially in the topsoil the plant cover could have a different influence on the nutrient contents, which is not reflected in the subsoil.

Table 12: Total contents of nutrients of all profiles of transect 1

Profile	Horizon	P	K	Ca	Mg	S	Na	Al	Fe	Mn
						[mg/kg]				
P1	Ah	869	19539	2550	8607	230	603	84295	43567	1113
	Ah-Bw1	780	20162	785	8381	198	615	89137	46902	1067
	Bw1	654	19749	559	8055	140	610	88340	47412	794
	Bw2	609	19486	569	8741	96	575	95962	52554	681
	Bw2-C	581	20677	492	8381	64	631	98395	52046	583
P2	Ah	623	6843	199	1582	221	235	89793	49415	325
	Bws-Ah	485	6749	138	1453	147	196	97299	53127	153
	Bws1	411	6842	103	1430	116	190	98433	54770	144
	Bws2	417	7590	109	1405	101	200	107005	56112	157
	Bws2-C	421	8171	104	1402	69	184	106387	55037	123
P3	Ah	622	17460	872	5079	251	827	79582	31826	944
	Ah-Bt1	641	17045	916	4549	263	784	72666	28828	842
	Bt1	607	16754	706	5771	224	703	94189	39457	826
	Bt2	682	17383	704	6019	180	749	97298	42138	800
P4	Ah	547	15273	172	5453	291	480	90590	43028	233
	Bw1	455	15335	148	6330	191	447	101531	43662	274
	2Bw2	449	19775	185	8385	130	514	113396	47426	371
	2Bw2-C	436	24138	156	11154	66	592	110827	44587	408

3.3.6 Effective CEC and base saturation

Topsoil

Comparing the effective cation exchange capacity (ECEC) of the topsoils of transect 1 P2 clearly had the lowest capacity (5.11 cmol$_c$/kg), while the ECEC value measured for the Ah horizon of P1 seems to be an outlier. With values between 5 and 7 cmol$_c$/kg the exchange capacities of the topsoils can be described as very low (Fig. 18a).

Clear differences were found regarding the base saturation (BS) in the topsoils. The values varied greatly. While the BS in the Ah horizon of P1 reached almost 100 %, BS of topsoil of P4 was around 7 %. These findings reflect the different coverage of exchangeable cations at the exchanger. In P1 Mg and Ca were the dominant cations involved in the ECEC (corresponding also with the pH of >6), whereas in P4 Al had the highest percentage of involved cations in the topsoil (Fig. 18b).

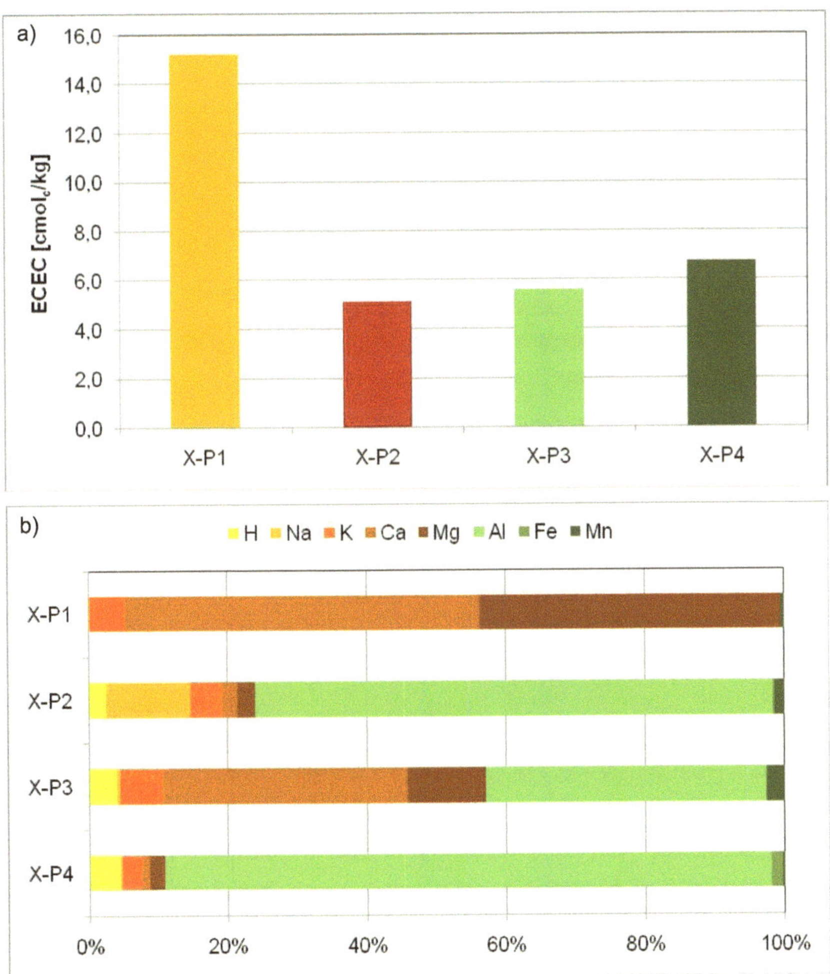

Figure 18: a) ECEC and b) the percentage of cations in Ah horizons of transect 1

Subsoil

In the subsoil a clear decreasing trend of ECEC was visible. Again P2 showed the lowest ECEC with values between 2.8 $cmol_c$/kg (Bws-Ah) and 0.9 $cmol_c$/kg (Bws2-C), while P1 had the highest values ranging from 7.2 $cmol_c$/kg (Ah-Bw1) to 3.7 $cmol_c$/kg (Bw2-C). Clear differences between P1, P3, and P4 could only be detected in the first subsoil horizons. With increasing depth the values of ECEC levelled (Fig. 19).

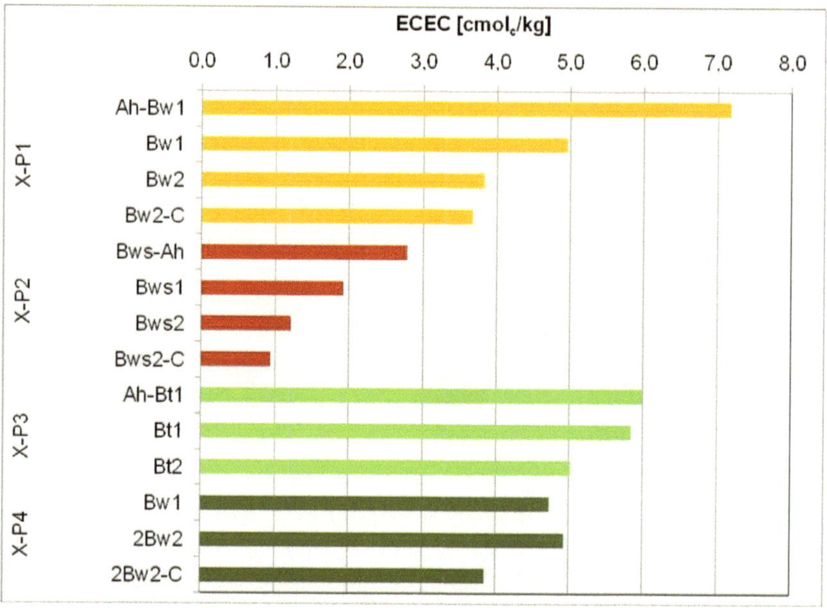

Figure 19: ECEC of subsoil horizons of transect 1

As already seen for the topsoil the base saturation showed also in the subsoil horizons a great variety with values between 2.9 % (P4, 2Bws-C) and 92.98 % (P1, Ah-Bw1). Fig. 20 shows the coverage of cations at the surface of the exchanger. Here again the dominant role of exchangeable Ca and Mg cations in Profile 1 can be seen, although their percentages are decreasing with increasing depth. Thus the percentage of Al cations is increasing in this profile. P2 and P4 showed similar to the topsoil findings the highest Al percentages throughout the whole profiles, reflecting again the comparable low BS of less than 20 %.

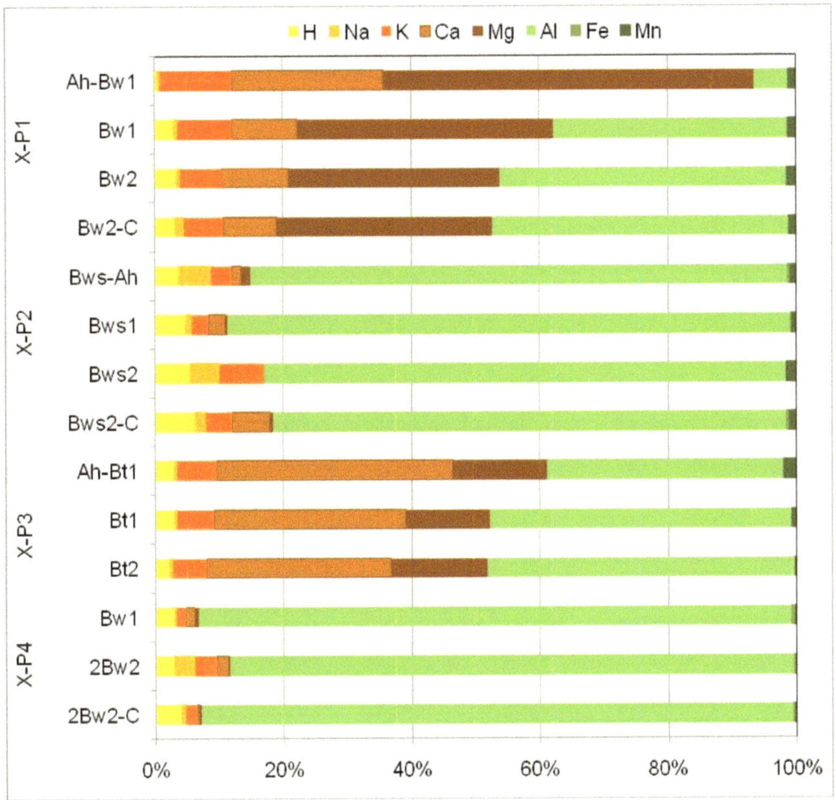

Figure 20: Percentage of exchangeable cations of subsoil horizons of transect 1

In general, effects of different altitudes could not be detected. A slight influence of land use effects may be seen for profile 2 (wasteland) compared to the other profiles of this transect (forests). But mainly different site conditions might have led to differentiations in the measured parameters between the profiles.

3.4 Transect 2

3.4.1 Depth and fine earth density

Topsoil

All profiles around Na Ban showed Ah horizon thicknesses of 10 to 20 cm, whereas the topsoil horizons of profiles 8 and 9 in Man Dian and Jiang Bian Zhan only made up to 4 to 7 cm.

The fine earth densities of topsoils showed a range from 0,96g/cm^3 (P5, Cha Chang forest) to 1.21 g/cm^3 (P7, Na Ban paddy and P8, Man Dian forest), without significant differences between the profiles.

Subsoil

Regarding the fine earth densities in the subsoil P5 and P7 showed clear increasing trends with increased depth, indicating a slight compaction of profile 7 due to agricultural use. The other profiles had only indistinct changes of the fine earth densities with the depth, although P8 shows the lowest density in the depth.

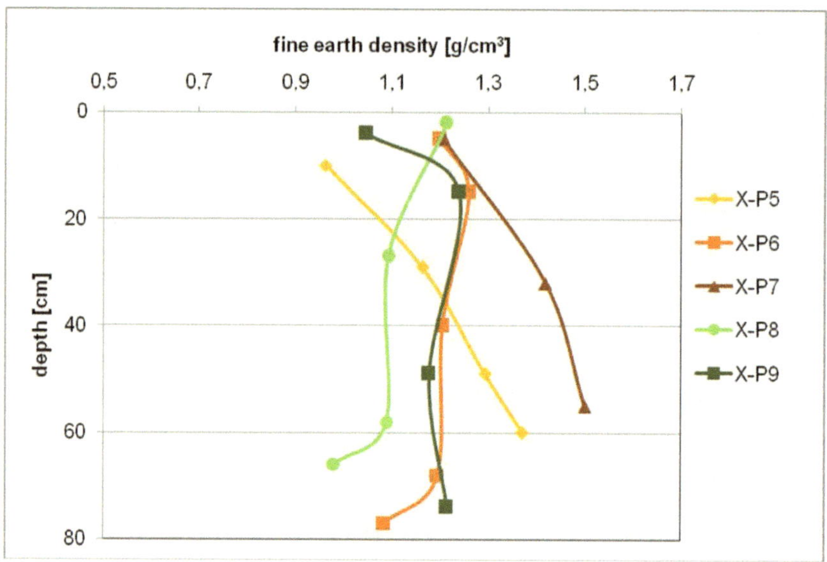

Figure 21: Fine earth densities all horizons of transect 2

3.4.2 Texture

Topsoil

On average, the profiles of transect 2 showed higher contents of silt (mean 37.90 %) compared to transect 1. The sand fraction had a proportion of about 24.15 % compared to 41.39 % in the Ah horizons of transect 1 (Fig. 21). Only in profile P7 (Na Ban, paddy field) sand reached a content of about 38.1 %. The textural classes were determined as clay (P5), silty clay (P6) and clay loam (P7, P8, P9).

Subsoil

In comparison with transect 2 all profiles of transect 2 showed increasing clay contents with depth from upper to lower horizons. Highest clay contents were found in the subsoil horizons of P6 (Na Ban, rubber stand; mean 55.99 %). On the other hand profile P7 had again the highest content of sand fractions (mean 39.7 %) and lowest content of clay (mean 28.04 %) as already observed for the topsoil. Consequently the dominating texture classes were clay and silty clay in P5, silty clay, clay and heavy clay in P6, loam and clay loam in P7; and clay loam and clay in P8 and P9 (Fig. 21).

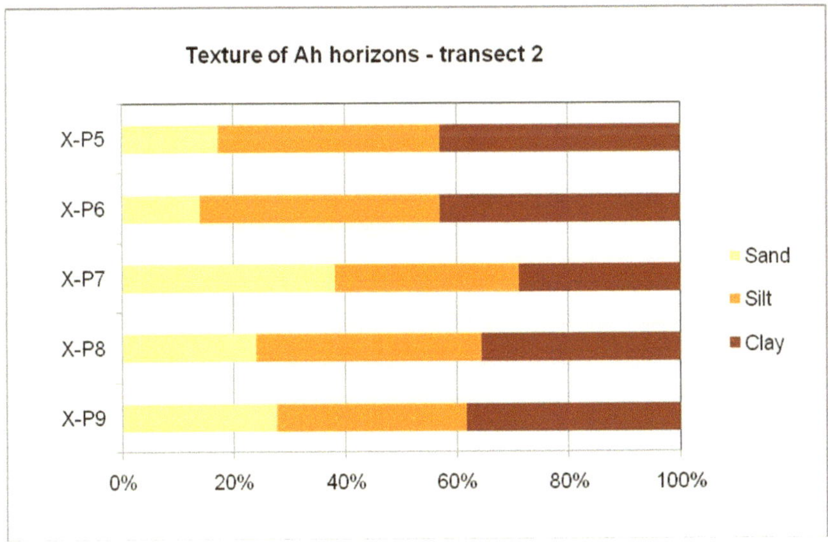

Figure 22: Distribution of soil particles in Ah horizons of transect 2

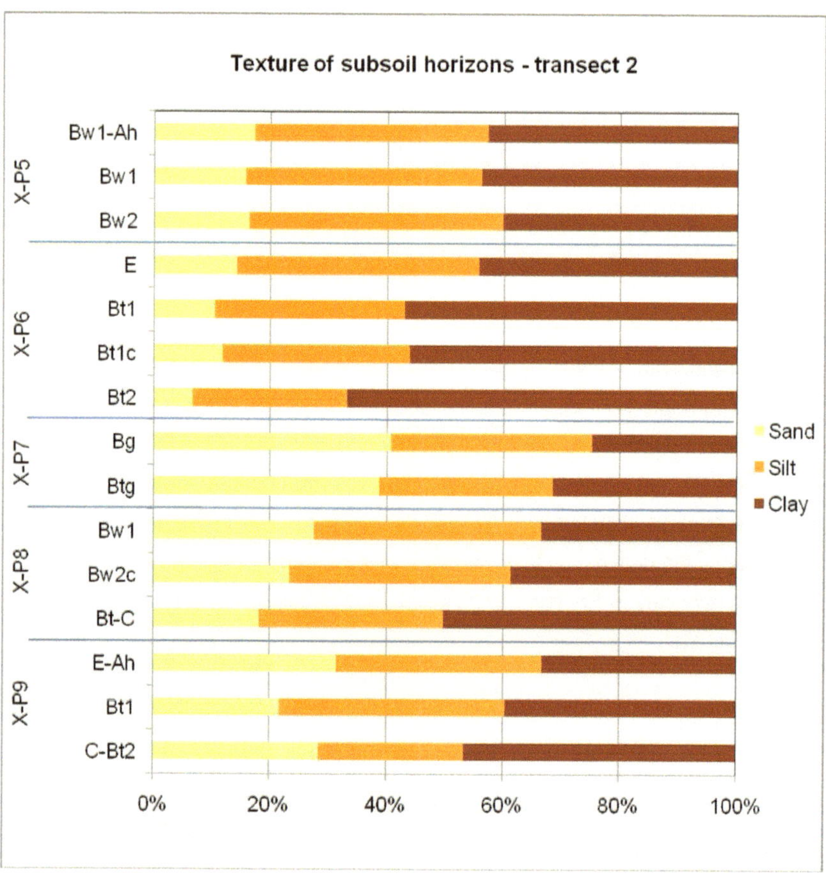

Figure 23: Distribution of soil particles in subsoil horizons of transect 2

3.4.3 pH value

Topsoil

The profiles P5, P6 and P9 were clearly showing more acid soil reactions in the Ah horizons (between 4.0 and 5.0) then P7 and P8 (pH above 5.0).

Subsoil

The subsoil of P7 had distinct higher ph values than the other profiles. This might show a land use effect. The farmer could have been applying alkaline substances in order to reach a higher pH value in the soil. As P7 was established in a paddy rice field in a river basin, the profile might be influenced by alluvial material causing a shift in the pH value (see also 3.4.5). The pH-values of all other profiles were nearly the same in the subsoil horizons and varied between pH 4.4 and pH 4.9 (Fig.24).

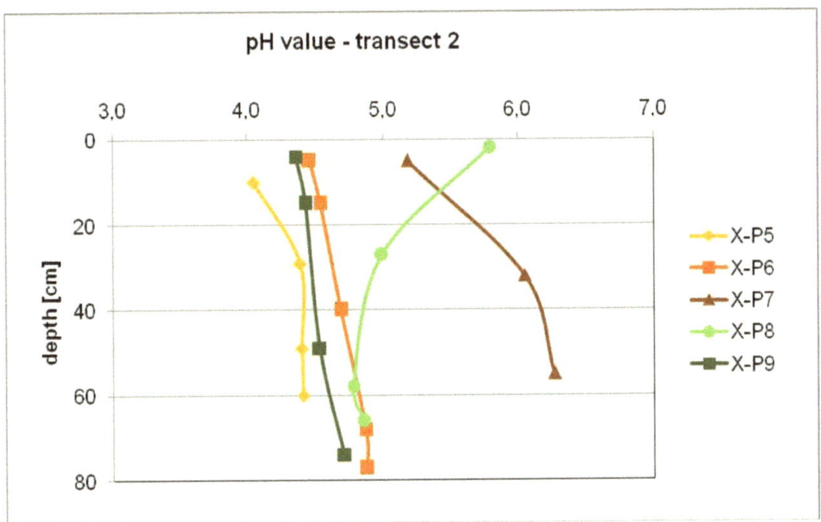

Figure 24: pH value of all horizons of transect 2

3.4.4 SOC and TN

Topsoil

The contents of soil organic carbon (SOC) in topsoils of the profiles near Na Ban (P5, P6, and P7) ranged from 12.4 g/kg at the paddy field (1.2 %, P7) to 19.4 g/kg at the forest site in Cha Chang (1.9 %, P5) (Fig. 25a). Profile 8 is a profile at the forest near Man Dian and shows a SOC content of 1.3 %, whereas the profile P9 under bamboo at Jiang Bian Zhan had the highest SOC content of more then 3 %. For the Profiles P5, P6 and P7 an influence of land use on the SOC content is visible. On the other hand the differences between the Na Ban site profiles compared to P8 and P9 could be a result of site effects.

The content of total nitrogen (TN) did show the same trends as seen for SOC, but in a smaller range (Fig. 25b). The C/N ratio in the topsoil ranged from 11.2 (P7) to 11.8 (P5) in Na Ban. Profile P9 on the other hand had a C/N ratio of 18.1 in the Ah horizon.

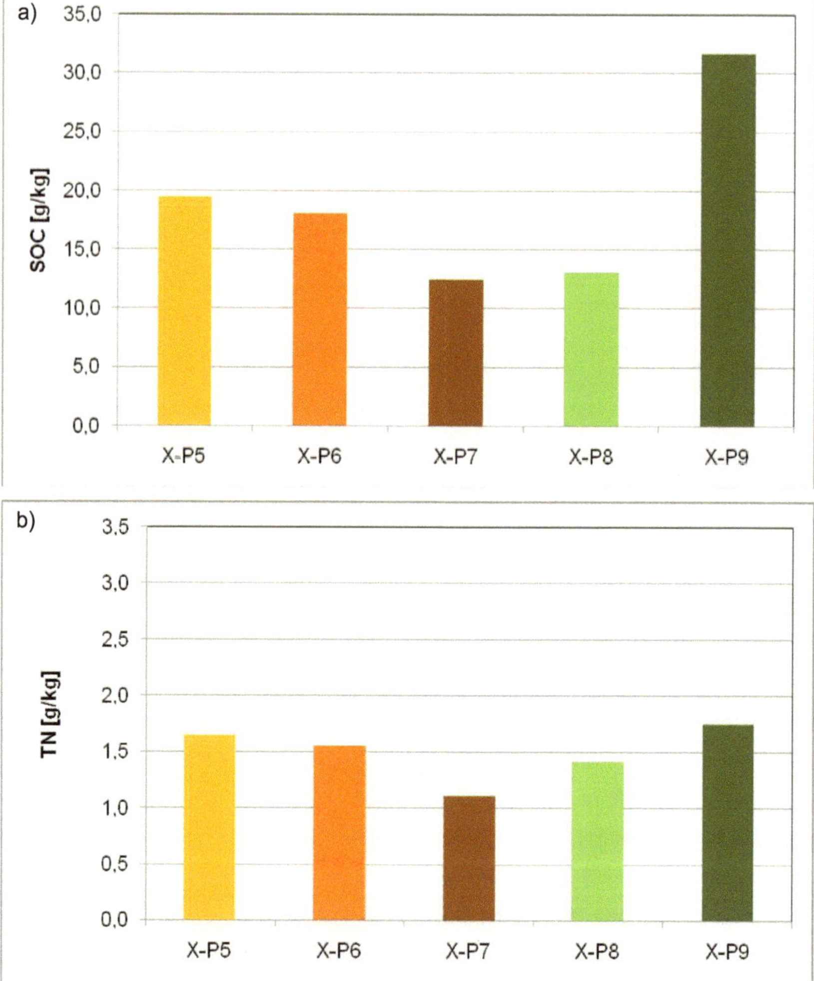

Figure 25: Contents of a) SOC and b) TN of Ah horizons of transect 2

Subsoil

In the subsoils the trends drawn for SOC contents were not as clear as in the topsoil (Fig 26). Here the contents of SOC were more or less the same except for the transitional horizon of P9 (E-Ah) showing a higher content of about 1.9 %.
In all profiles the SOC contents decreased with increasing depth.

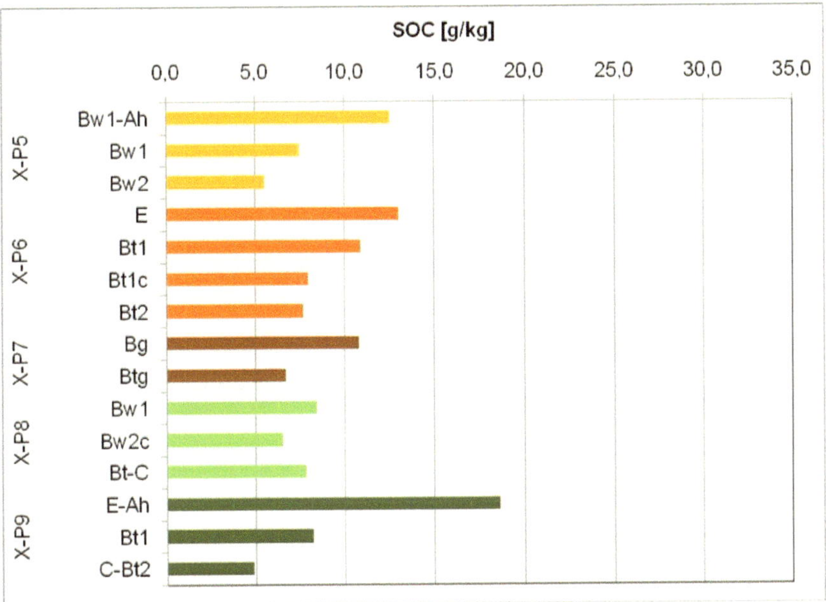

Figure 26: SOC content of subsoil horizons of transect 2

The TN contents of subsoil horizons did not show clear trends and ranged between 0.6 g/kg (P5 Bw2) and 1.4 g/kg (P9, Ah-Bt1) (Fig. 27). Again the most obvious differences in the TN contents were found in the transitional horizon of P9 showing the highest content with 1.4 g/kg at the Ah-Bt1 horizon.

The SOC/TN ratios decreased from 11.4 (mean of first transitional horizons) to 8.5 (lowest horizons) in the subsoil horizons and did not change significantly between the profiles.

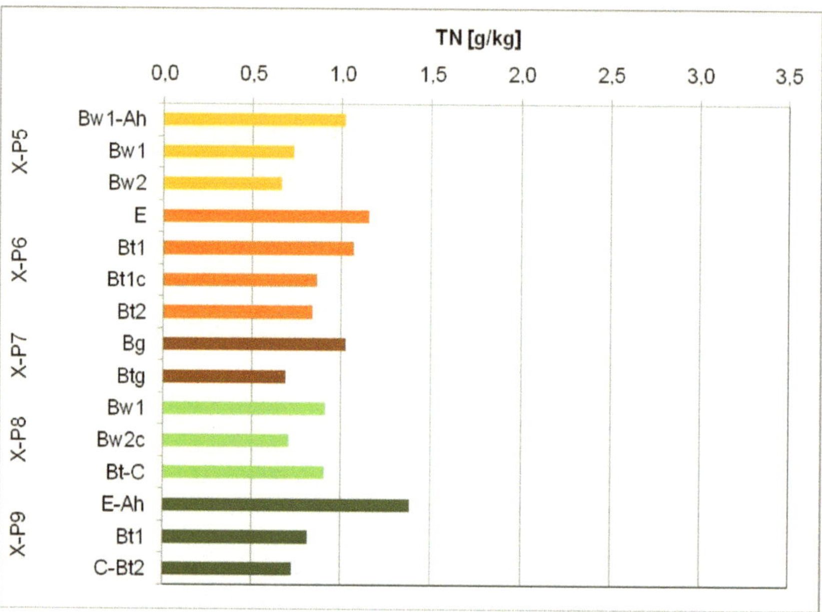

Figure 27: TN content of subsoil horizons of transect 2

3.4.5 Nutrients

Topsoil

The total contents of nutrients are shown in Tab. 13. Regarding the Ah horizons the highest total contents of calcium and magnesium were measured for the Ah horizon of P7. An application of alkaline substances or a natural influence of alluvial material was already discussed (see 3.4.3). The highest content of Mn was found in P8.

Subsoil

Comparing the whole profiles again P7 shows the highest contents of Ca and Mg, whereas P8 again had the highest Mn contents. The elements Al and Fe were slightly enriched in profile P6, whereas P7 clearly showed lower contents of Al compared to all other profiles. Significant differences were also found for sodium. Here P7 and P9 had the highest contents

Table 13: Total contents of nutrients of all profiles of transect 2

Profile	Horizon	P	K	Ca	Mg	S	Na	Al	Fe	Mn
						[mg/kg]				
P5	**Ah-Bw1**	241	18363	146	2304	138	570	90347	25893	28
	Bw1-Ah	217	18996	142	2322	74	588	93523	28121	28
	Bw1	188	18990	141	2456	62	583	98114	28668	28
	Bw2	167	19452	191	2348	18	614	96487	28651	30
P6	**Ah-E**	607	15326	268	3021	179	477	84463	41801	445
	E	578	16145	254	3147	161	507	92329	43752	358
	Bt1	597	17455	224	3477	116	520	106387	53903	267
	Bt1c	622	18466	238	3683	63	597	113379	59267	227
	Bt2	609	17179	217	3223	129	543	100535	57753	214
P7	**Apg**	473	14848	1739	3339	167	1390	45602	28462	244
	Bg	486	16065	2407	5710	163	1295	61785	36661	564
	Btg	435	16511	2279	6608	159	1195	67527	41121	757
P8	**Ah**	663	13492	634	3743	215	360	65708	39254	844
	Bw1	569	13260	251	3406	168	328	66352	39536	835
	Bw2c	564	13950	159	4188	172	329	81700	46557	755
	Bt-C	604	13496	187	4356	163	336	88109	53277	578
P9	**Ah**	314	13086	233	859	163	776	48226	31362	129
	E-Ah	322	15112	213	943	141	867	55441	35901	119
	Bt1	323	18423	272	1433	58	1045	87614	48374	114
	C-Bt2	292	19837	211	1534	54	1144	98728	54728	125

3.4.6 Effective CEC and base saturation

Topsoil

The effective cation exchange capacities (ECEC) of the topsoils of transect 2 showed values between 5.2 cmol$_c$/kg (P6) and 6.2 cmol$_c$/kg (P5 and P8). Significant differences were not detected (Fig. 28a).

Similar to transect 1 the base saturation (BS) in the topsoils of transect 2 varied greatly. The BS in the Ah horizon of P5 and P9 reached only 8.6 % and 10.8 %, while for P7 and P8 a BS of 91.4 % and 97.9 % was measured. As already discussed these findings reflect the different coverage of exchangeable cations at the exchanger. In P7 and P8 Ca and Mg played a dominant role in the ECEC (corresponding also with the pH of >5), whereas in the other profiles Al showed the highest percentage (Fig. 28b).

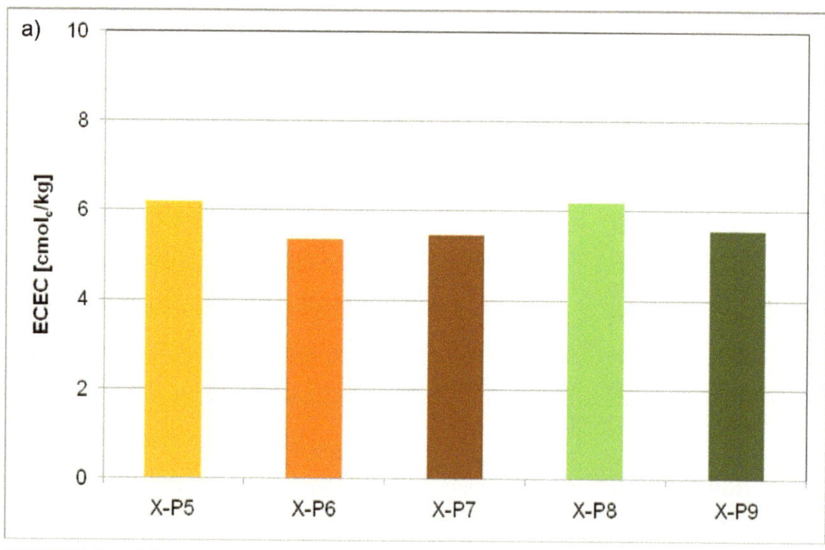

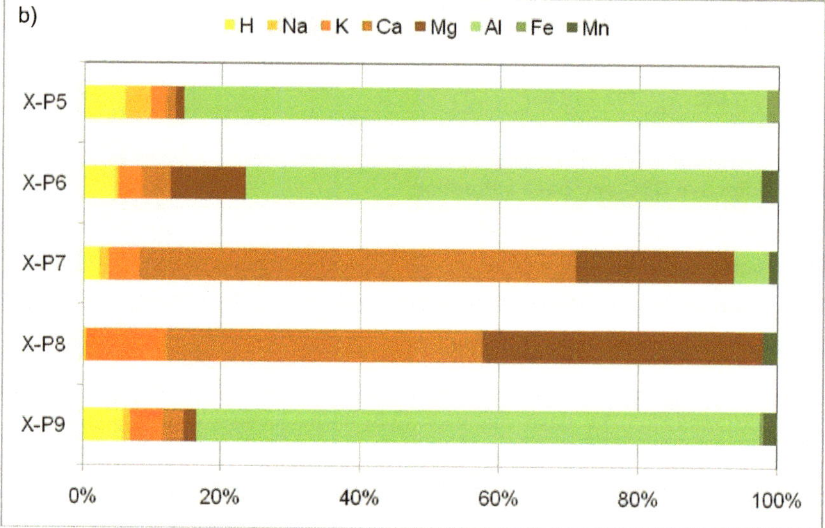

Figure 28: a) ECEC and b) the percentage of cations in Ah horizons of transect 2

Subsoil

In the subsoil P7 showed the highest effective cation exchange capacity (7.9 cmol$_c$/kg in Bg; 9.3 cmol$_c$/kg in Btg). A clear decreasing trend of ECEC was visible only in P6, whereas P7 and P9 showed clear increasing trends. The lowest ECEC in subsoils was found in P8 with values between 3.3 cmol$_c$/kg (Bw2c) and 3.4 cmol$_c$/kg (Bw1) (Fig. 29). Thus, a clear influence of land use effects can be determined comparing profile P5, P6 and P7, while for P9 again site conditions may be the reason for the differences in ECEC values compared to all other profiles.

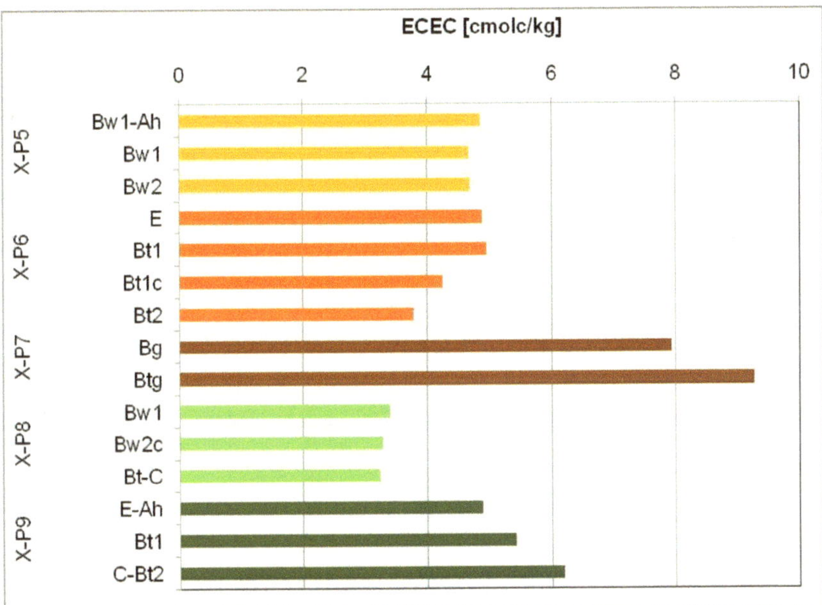

Figure 29: ECEC of subsoil horizons of transect 2

The base saturation again varied significantly reflecting the same trends as seen in the topsoil. The values ranged from 2.2 % (P9, C-Bt2) to 99.44 % (P7, Bg). The coverage of cations at the surface of the exchanger is shown in the following figure (Fig. 30). Correspondent to the high BS values in P7 the dominant role of exchangeable Ca and Mg cations in this profile can be seen. In the subsoils of P5, P6 and P9 Al cations were dominant.

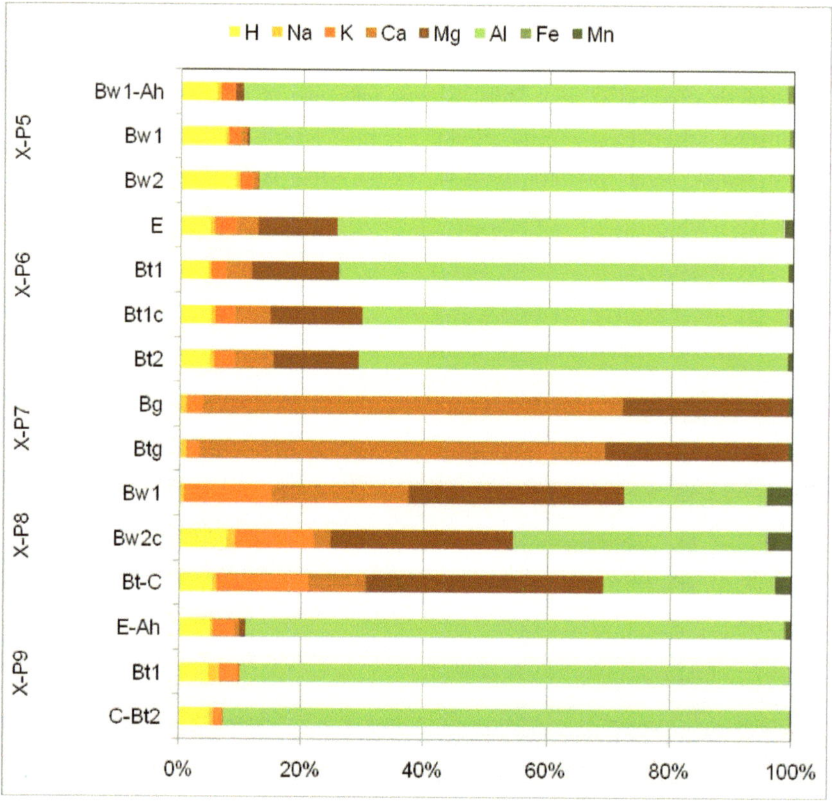

Figure 30: Percentage of exchangeable cations of subsoil horizons of transect 1

In summary, influences of land use effects could be seen for profile P5, P6 and P7, whereas P8 and P9 showed differences to the other profiles due to there different site conditions in Man Dian and Jiang Bian Zhan, although influences of the vegetation cover (forest, bamboo forest) might also interfere.

4 Literature

ACHILLES (1997): Agroecological Data on Ethnic Communities in Selected Watersheds of Xishuangbanna, Yunnan, China, Bangkok.

AD-HOC-AG BODEN (2005): Bodenkundliche Kartieranleitung, 5. Auflage, Hannover.

ANONYMUS (1987): Xishuangbanna Ziranbaohuqu. – Yunnan: Verlag der Wissenschaft und Technik, 541 p. In APEL, U. (1996): Der Dorfwald von Moxie, Traditio-nelle Dorfwaldbewirtschaftung in Xishuangbanna, Südwest-China, Göttingen.

APEL, U. (1996): Der Dorfwald von Moxie, Traditionelle Dorfwaldbewirtschaftung in Xishuangbanna, Südwest-China, Göttingen.

BRUENIG, E.F. (ED.) (1986): Ecologic-Socioeconomic System: Analysis and Simulation; A Guide for Application of System analysis to the Conversation, Utilization and Development of Tropical and Subtropical Land Resources in China, Hamburg.

FAO/UNESCO (1988): Soil Map of the World, Rome.

FAO/UNESCO (1998): World Reference Base for Soil Resources 1998, Rome.

FAO/UNESCO (2006a): Guidelines for Soil Description, Fourth Edition, Rome.

FAO/UNESCO (2006b): World Reference Base for Soil Resources 2006, Rome.

HE, Z.L.; ZHANG, M.K. and WILSON, M.J. (2004a): Distribution and Classification of Red Soil in China. Pp. 29-33 in WILSON, M.J.; HE, ZL und YANG, XE (EDS.) (2004): The Red Soil of China - Their Nature, Management and Utilization. Kluwer Academic Publishers, Dordrecht.

HE, Z.L.; ZHANG, M.K. and WILSON, M.J. (2004b): Chemical Soil Constraints to Crop Production on Chinese Red Soils. Pp. 103-110 in WILSON, M.J.; HE, ZL and YANG, XE (EDS.) (2004): The Red Soil of China-Their Nature, Management and Utilization. Kluwer Academic Publishers, Dordrecht.

SCHEFFER, F. & SCHACHTSCHABEL, P. (1998): Lehrbuch der Bodenkunde, 14. Auflage, Ferdinand Enke Verlag Stuttgart.

SHI, X.Z.; YU, D.S.; WARNER, E.D.; SUN, W.X.; PETERSEN, G.W.; GONG, Z.T. and LIN, H. (2006): Cross-Reference System for Translating Between Genetic Soil Classification of China and Soil Taxonomy. SSSA Journal 70, 78-83.

SPAARGAREN, O.C. and DECKERS, J. (1998): The World Reference Base for Soil Resources: An Introduction with Special Reference to Soils of Tropical Forest Ecosystems. Pp. 21-28 in SCHULTE, A. and RUHIYAT, D. (Eds.): Soils of Tropical Forest Ecosystems: Characteristics, Ecology and Management. Springer-Verlag Berlin Heidelberg.

WANG, J.R. (2000): Indigenous Knowledge in Upland Farming Management for Community Development Strategies in Xishuangbanna, Southern Yunnan, China. Pp. 303-308 in XU, J.C. (ED.): Links between Cultures and Biodiversity: Proceedings of the Cultures and Biodiversity Congress 2000, 20-30 July, Yunnan, P.R. China. Yunnan Science and Technology Press.

WILSON, M.J.; HE, Z.L. and YANG, X.E. (EDS.) (2004): The Red Soils of China-Their Nature, Management and Utilization. Kluwer Academic Publishers, Dordrecht.

WU, Z. and OU, X. (1995): The Xishuangbanna Biosphere Reserve – A Tropical Land of Natural and Cultural Diversity (China). South-South Cooperation Programme on Environmentally Sound Socio-Economic Development in the Humid Tropics, Working Papers, No. 2, 1995. UNESCO, The Man and the Biosphere Programme.

YUNNAN SOCIETY OF ECOLOGICAL ECONOMICS, XISHUANGBANNA NATURE RESERVE ADMINISTRATION & YUNNAN FORESTRY INVESTIGATION AND PLANNING INSTITUTE (1992): Xishuangbanna – A Nature Reserve of China. – Beijing: China Forestry Publishing House.

ZHANG, K.Y. (1986): The Influence of Deforestation of Tropical Rainforests on Local Climate and Disaster in the Xishuangbanna Region of China. – Climatological Notes 35, 223-236.

5 Annex

Table A 1: Coordinates of the profiles

Profile	Site	N	E	Elevation [m a.s.l.]	Aspect	Land use	Date of description
Transect 1							
P1	Hui Lao Xin Zhai	N 22°07.652'	E 100°38.849'	approx. 1000	SWW	forest	2009-11-08
P2	Gei Yang Gong Di	N 22°09.543'	E 100°37.485'	1664	SE	waste land	2009-11-08
P3	Beng Gang Lahu	N 22°09.459'	E 100°33.953'	1560	NNW	forest	2009-11-07
P4	Bang Gang Ha Ni	N 22°07.091'	E 100°35.114'	1764	WW	forest	2009-11-09
Transect 2							
P5	Cha Chang	N 22°09.479'	E 100°39.911'	689	SS	forest	2009-11-10
P6	Naban, rubber old	N 22°10.418'	E 100°39.494'	683	NNE	rubber	2009-11-10
P7	Naban, paddy field	N 22°10.013'	E 100°39.696'	659	NN	paddy	2009-11-11
P8	Man Dian	N 22°07.763'	E 100°39.956'	686	NE	forest	2009-11-13
P9	Jiang Bian Zhan	N 22°70.804'	E 100°43.055'	703	SS	bamboo	2009-11-12

Table A 2: Data of profile 1 (Hui Lao Xin Zhai)

Soil physics / **C and N**

Profile	Depth	BD fresh [g/cm³]	BD air dry [g/cm³]	BD oven dry [g/cm³]	Sand [%]	Silt [%]	Clay [%]	Texture	pH (H2O)	SOC [g/kg]	TN [g/kg]	C/N	SOC [t/ha]	TN [t/ha]
X-P1	0-8	1.25	1.04	1.02	46.26	25.34	28.39	SCL	6.29	34.59	2.37	14.61	28.12	1.92
X-P1	8-26	1.43	1.19	1.17	43.39	24.15	32.44	CL	5.60	19.45	1.59	12.23	40.87	3.34
X-P1	26-48	1.54	1.28	1.26	48.84	22.17	28.99	SCL	5.28	11.74	1.00	11.77	32.43	2.75
X-P1	48-78	1.51	1.26	1.24	49.57	23.01	27.42	SCL	5.27	8.00	0.74	10.77	29.68	2.76
X-P1	78-	1.49	1.25	1.23	55.79	22.66	24.91	SCL	5.37	5.77	0.56	10.37	15.61	1.51

Nutrients

Profile	Depth	P [mg/kg]	K [mg/kg]	Ca [mg/kg]	Mg [mg/kg]	S [mg/kg]	Na [mg/kg]	Al [mg/kg]	Fe [mg/kg]	Mn [mg/kg]	Zn [mg/kg]
X-P1	0-8	869	19539	2550	8607	230	603	84295	43567	1113	146
X-P1	8-26	780	20162	785	8381	198	615	89137	46902	1067	118
X-P1	26-48	654	19749	559	8055	140	610	88340	47412	794	105
X-P1	48-78	609	19486	569	8741	96	575	95962	52554	681	103
X-P1	78-	581	20677	492	8381	64	631	98395	52046	583	101

ECEC

Profile	Depth	ECEC [cmolc/kg]	BS [%]	H [µeq/g]	Na [µeq/g]	K [µeq/g]	Ca [µeq/g]	Mg [µeq/g]	Al [µeq/g]	Fe [µeq/g]	Mn [µeq/g]
X-P1	0-8	15.20	99.53	0.00	0.16	7.70	77.62	65.79	0.00	0.01	0.70
X-P1	8-26	7.17	92.98	0.32	0.25	8.13	16.82	41.44	3.79	0.02	0.90
X-P1	26-48	4.95	59.25	1.50	0.23	4.27	5.02	19.82	18.01	0.03	0.64
X-P1	48-78	3.82	50.37	1.30	0.20	2.52	3.98	12.57	17.12	0.04	0.52
X-P1	78-	3.66	49.33	1.15	0.49	2.31	3.01	12.26	16.96	0.04	0.42

Table A 3 Data of profile 2 (Gei Yang Gong Di)

Soil physics

C and N

Profile	Depth	BD [g/cm³]	BD air dry [g/cm³]	BD oven dry [g/cm³]	Sand [%]	Silt [%]	Clay [%]	Texture	pH (H2O)	SOC [g/kg]	TN [g/kg]	C/N	SOC [t/ha]	TN [t/ha]
X-P2	0-15	1.13	0.92	0.90	41.67	18.73	39.26	CL	4.81	42.80	2.38	17.97	57.56	3.20
X-P2	15-25	1.43	1.16	1.14	39.77	16.97	42.44	C	4.83	21.05	1.23	17.12	23.96	1.40
X-P2	25-43	1.61	1.31	1.29	36.09	15.56	48.32	C	4.90	13.38	0.80	16.64	31.03	1.86
X-P2	43-70	1.60	1.30	1.29	36.57	17.26	46.16	C	5.14	8.62	0.56	15.40	35.52	2.31
X-P2	70-	1.56	1.32	1.30	42.97	21.34	35.67	CL	5.26	4.07	0.31	13.35	13.27	0.99

Nutrients

Profile	Depth	P [mg/kg]	K [mg/kg]	Ca [mg/kg]	Mg [mg/kg]	S [mg/kg]	Na [mg/kg]	Al [mg/kg]	Fe [mg/kg]	Mn [mg/kg]	Zn [mg/kg]
X-P2	0-15	623	6843	199	1582	221	235	89793	49415	325	37
X-P2	15-25	485	6749	138	1453	147	196	97299	53127	153	36
X-P2	25-43	411	6842	103	1430	116	190	98433	54770	144	33
X-P2	43-75	417	7590	109	1405	101	200	107005	56112	157	35
X-P2	75-	421	8171	104	1402	69	184	106387	55037	123	32

ECEC

Profile	Depth	ECEC [cmolc/kg]	BS [%]	H [µeq/g]	Na [µeq/g]	K [µeq/g]	Ca [µeq/g]	Mg [µeq/g]	Al [µeq/g]	Fe [µeq/g]	Mn [µeq/g]
X-P2	0-15	5.11	21.57	1.20	6.28	2.28	1.18	1.29	38.00	0.13	0.76
X-P2	15-25	2.78	11.26	1.03	1.38	0.98	0.38	0.39	23.25	0.11	0.26
X-P2	25-43	1.92	6.52	0.92	0.19	0.52	0.49	0.05	16.85	0.05	0.13
X-P2	43-75	1.20	11.51	0.66	0.56	0.82	0.00	0.00	9.78	0.03	0.17
X-P2	75-	0.93	11.99	0.59	0.15	0.40	0.52	0.04	7.48	0.04	0.10

Table A 4: Data of profile 3 (Beng Gang Lahu)

Soil physics / C and N

Profile	Depth	BD	BD air dry	BD oven dry	Sand	Silt	Clay	Texture	pH	SOC	TN	C/N	SOC	TN
		[g/cm³]	[g/cm³]	[g/cm³]	[%]	[%]	[%]		(H2O)	[g/kg]	[g/kg]		[t/ha]	[t/ha]
X-P3	0-15	1.56	1.18	1.16	44.04	21.70	34.22	CL	4.75	20.64	2.12	9.73	35.81	3.68
X-P3	15-23	1.56	1.21	1.19	40.82	22.32	36.84	CL	4.89	32.07	2.96	10.84	30.46	2.81
X-P3	23-50	1.43	1.07	1.05	35.73	17.86	46.39	C	5.03	15.07	1.70	8.84	42.83	4.84
X-P3	50-	1.56	1.17	1.14	38.65	19.53	42.64	C	5.17	11.86	1.36	8.70	67.86	7.80

Nutrients

Profile	Depth	P	K	Ca	Mg	S	Na	Al	Fe	Mn	Zn
		[mg/kg]	[mg/kg]	[mg/kg]	[mg/kg]	[mg/kg]	[mg/kg]	[mg/kg]	[mg/kg]	[mg/kg]	[mg/kg]
X-P3	0-15	622	17460	872	5079	251	827	79582	31826	944	91
X-P3	15-23	641	17045	916	4549	263	784	72666	28828	842	72
X-P3	23-50	607	16754	706	5771	224	703	94189	39457	826	99
X-P3	50-	682	17383	704	6019	180	749	97298	42138	800	105

ECEC

Profile	Depth	ECEC	BS	H	Na	K	Ca	Mg	Al	Fe	Mn
		[cmolc/kg]	[%]	[µeq/g]	[µeq/g]	[µeq/g]	[µeq/g]	[µeq/g]	[µeq/g]	[µeq/g]	[µeq/g]
X-P3	0-15	5.60	53.09	2.16	0.29	3.49	19.72	6.22	22.56	0.08	1.45
X-P3	15-23	5.99	58.26	1.76	0.26	3.75	22.02	8.85	22.00	0.07	1.17
X-P3	23-50	5.84	49.11	1.72	0.31	3.38	17.41	7.58	27.55	0.05	0.39
X-P3	50-	5.02	49.43	1.12	0.28	2.63	14.42	7.47	24.08	0.04	0.13

Table A 5: Data of profile 4 (Beng Gang Hani)

Soil physics

									C and N					
Profile	Depth	BD [g/cm³]	BD air dry [g/cm³]	BD oven dry [g/cm³]	Sand [%]	Silt [%]	Clay [%]	Texture	pH (H2O)	SOC [g/kg]	TN [g/kg]	C/N	SOC [t/ha]	TN [t/ha]
X-P4	0-14	1.37	1.01	0.99	33.28	19.02	47.68	C	4.06	31.51	2.81	11.23	43.46	3.87
X-P4	14-46	1.42	1.08	1.06	38.18	18.84	42.97	C	4.39	16.38	1.26	13.02	55.41	4.26
X-P4	46-96	1.54	1.23	1.22	45.70	23.39	30.89	SCL	4.64	6.40	0.45	14.13	38.96	2.76
X-P4	96-	1.44	1.20	1.18	61.61	22.59	15.79	SL	4.79	2.74	0.21	13.34	1.30	0.10

Nutrients

Profile	Depth	P [mg/kg]	K [mg/kg]	Ca [mg/kg]	Mg [mg/kg]	S [mg/kg]	Na [mg/kg]	Al [mg/kg]	Fe [mg/kg]	Mn [mg/kg]	Zn [mg/kg]
X-P4	0-14	541	15273	172	5453	291	480	90590	43028	233	57
X-P4	14-46	455	15335	148	6330	191	447	101531	43662	274	70
X-P4	46-96	449	19775	185	8385	130	514	113396	47426	371	85
X-P4	96-	436	24138	156	11154	66	592	110827	44587	408	90

ECEC

Profile	Depth	ECEC [cmolc/kg]	BS [%]	H [µeq/g]	Na [µeq/g]	K [µeq/g]	Ca [µeq/g]	Mg [µeq/g]	Al [µeq/g]	Fe [µeq/g]	Mn [µeq/g]
X-P4	0-14	6.74	6.48	2.93	0.18	2.02	0.73	1.44	58.76	1.06	0.28
X-P4	14-46	4.73	3.82	1.39	0.13	0.76	0.65	0.26	43.70	0.29	0.10
X-P4	46-96	4.93	8.62	1.46	1.56	1.78	0.87	0.04	43.43	0.09	0.08
X-P4	96-	3.85	2.92	1.59	0.22	0.70	0.12	0.07	35.55	0.13	0.09

Table A 6: Data of profile 5 (Cha Chang)

Soil physics / C and N

Profile	Depth	BD [g/cm³]	BD air dry [g/cm³]	BD oven dry [g/cm³]	Sand [%]	Silt [%]	Clay [%]	Texture	pH (H2O)	SOC [g/kg]	TN [g/kg]	C/N	SOC [t/ha]	TN [t/ha]
X-P5	0-20	1.20	0.98	0.96	17.26	39.64	43.06	C	4.05	19.39	1.65	11.75	37.38	3.18
X-P5	20-38	1.44	1.18	1.16	17.42	39.83	42.68	C	4.39	12.49	1.02	12.26	26.19	2.14
X-P5	38-60	1.59	1.31	1.30	15.98	40.29	43.66	SiC	4.40	7.48	0.73	10.23	21.30	2.08
X-P5	60-	1.66	1.39	1.37	16.52	43.40	40.08	SiC	4.41	5.52	0.66	8.32	30.29	3.64

Nutrients

Profile	Depth	P [mg/kg]	K [mg/kg]	Ca [mg/kg]	Mg [mg/kg]	S [mg/kg]	Na [mg/kg]	Al [mg/kg]	Fe [mg/kg]	Mn [mg/kg]	Zn [mg/kg]
X-P5	0-20	241	18363	146	2304	138	570	90347	25893	28	18
X-P5	20-38	217	18996	142	2322	74	588	93523	28121	28	19
X-P5	38-60	188	18990	141	2456	62	583	98114	28668	28	16
X-P5	60-	167	19452	191	2348	18	614	96487	28651	30	16

ECEC

Profile	Depth	ECEC [cmolc/kg]	BS [%]	H [µeq/g]	Na [µeq/g]	K [µeq/g]	Ca [µeq/g]	Mg [µeq/g]	Al [µeq/g]	Fe [µeq/g]	Mn [µeq/g]
X-P5	0-20	6.18	8.56	3.68	2.30	1.37	0.83	0.79	51.72	1.04	0.04
X-P5	20-38	4.86	4.36	2.83	0.37	1.11	0.00	0.63	43.12	0.43	0.06
X-P5	38-60	4.66	3.82	3.41	0.26	0.78	0.57	0.18	41.16	0.22	0.07
X-P5	60-	4.68	3.74	4.18	0.39	0.83	0.50	0.03	40.63	0.19	0.05

Table A 7: Data of profile 6 (Na Ban, rubber old)

Soil physics | | | | | | | | **C and N**

Profile	Depth	BD [g/cm³]	BD air dry [g/cm³]	BD oven dry [g/cm³]	Sand [%]	Silt [%]	Clay [%]	Texture	pH (H2O)	SOC [g/kg]	TN [g/kg]	C/N	SOC [t/ha]	TN [t/ha]
X-P6	0-10	1.56	1.22	1.20	13.97	42.82	43.13	SiC	4.46	18.06	1.55	11.66	21.67	1.86
X-P6	10-20	1.61	1.28	1.26	14.39	41.33	44.28	SiC	4.54	13.00	1.15	11.30	16.39	1.45
X-P6	20-60	1.60	1.23	1.21	10.62	32.34	57.00	C	4.69	10.89	1.07	10.18	52.54	5.16
X-P6	60-77	1.62	1.22	1.20	11.97	31.90	55.91	C	4.87	7.97	0.86	9.23	16.21	1.76
X-P6	77-	1.50	1.10	1.09	6.73	26.43	66.76	HC	4.88	7.68	0.84	9.18	19.16	2.09

Nutrients

Profile	Depth	P [mg/kg]	K [mg/kg]	Ca [mg/kg]	Mg [mg/kg]	S [mg/kg]	Na [mg/kg]	Al [mg/kg]	Fe [mg/kg]	Mn [mg/kg]	Zn [mg/kg]
X-P6	0-10	607	15326	268	3021	179	477	84463	41801	445	84
X-P6	10-20	578	16145	254	3147	161	507	92329	43752	358	86
X-P6	20-60	597	17455	224	3477	116	520	106387	53903	267	94
X-P6	60-77	622	18466	238	3683	63	597	113379	59267	227	96
X-P6	77-	609	17179	217	3223	129	543	100535	57753	214	102

ECEC

Profile	Depth	ECEC [cmolc/kg]	BS [%]	H [µeq/g]	Na [µeq/g]	K [µeq/g]	Ca [µeq/g]	Mg [µeq/g]	Al [µeq/g]	Fe [µeq/g]	Mn [µeq/g]
X-P6	0-10	5.36	18.97	2.42	0.25	1.79	2.31	5.81	39.69	0.05	1.28
X-P6	10-20	4.88	20.68	2.37	0.38	1.62	1.80	6.30	35.55	0.05	0.74
X-P6	20-60	4.94	21.28	2.29	0.20	1.20	2.05	7.07	36.22	0.04	0.36
X-P6	60-77	4.24	24.69	2.11	0.34	1.31	2.42	6.41	29.53	0.04	0.28
X-P6	77-	3.78	24.19	1.86	0.24	1.24	2.39	5.28	26.46	0.03	0.28

Table A 8: Data of profile 7 (Na Ban, paddy field)

Soil physics   **C and N**

Profile	Depth	BD [g/cm³]	BD air dry [g/cm³]	BD oven dry [g/cm³]	Sand [%]	Silt [%]	Clay [%]	Texture	pH (H2O)	SOC [g/kg]	TN [g/kg]	C/N	SOC [t/ha]	TN [t/ha]
X-P7	0-10	1.65	1.22	1.21	38.01	32.91	29.00	CL	5.18	12.42	1.11	11.18	15.01	1.34
X-P7	10-55	1.78	1.44	1.42	40.68	34.69	24.63	L	6.05	10.82	1.02	10.56	53.75	5.09
X-P7	55-	1.90	1.52	1.50	38.78	29.75	31.46	CL	6.27	6.68	0.69	9.74	55.18	5.67

Nutrients

Profile	Depth	P [mg/kg]	K [mg/kg]	Ca [mg/kg]	Mg [mg/kg]	S [mg/kg]	Na [mg/kg]	Al [mg/kg]	Fe [mg/kg]	Mn [mg/kg]	Zn [mg/kg]
X-P7	0-10	473	14848	1739	3339	167	1390	45602	28462	244	61
X-P7	10-55	486	16065	2407	5710	163	1295	61785	36661	564	62
X-P7	55-	435	16511	2279	6608	159	1195	67527	41121	757	66

ECEC

Profile	Depth	ECEC [cmolc/kg]	BS [%]	H [µeq/g]	Na [µeq/g]	K [µeq/g]	Ca [µeq/g]	Mg [µeq/g]	Al [µeq/g]	Fe [µeq/g]	Mn [µeq/g]
X-P7	0-10	5.45	91.39	1.21	0.75	2.42	34.31	12.36	2.76	0.10	0.62
X-P7	10-55	7.94	99.44	0.00	0.82	2.01	54.45	21.70	0.00	0.02	0.42
X-P7	55-	9.26	99.41	0.00	0.91	1.88	61.29	27.94	0.00	0.07	0.48

Table A 9: Data of profile 8 (Man Dian)

Soil physics

									C and N					
Profile	Depth	BD	BD air dry	BD oven dry	Sand	Silt	Clay	Texture	pH	SOC	TN	C/N	SOC	TN
		[g/cm³]	[g/cm³]	[g/cm³]	[%]	[%]	[%]		(H2O)	[g/kg]	[g/kg]		[t/ha]	[t/ha]
X-P8	0-4	1.60	1.23	1.21	23.90	40.20	35.73	CL	5.79	13.03	1.41	9.23	6.33	0.69
X-P8	4-50	1.42	1.11	1.09	27.63	38.94	33.27	CL	4.99	8.42	0.91	9.24	42.35	4.58
X-P8	50-66	1.42	1.10	1.09	23.43	37.90	38.57	CL	4.78	6.48	0.71	9.20	11.31	1.23
X-P8	66-	1.33	1.00	0.98	18.33	31.38	50.04	C	4.86	7.80	0.91	8.59	26.04	3.03

Nutrients

Profile	Depth	P [mg/kg]	K [mg/kg]	Ca [mg/kg]	Mg [mg/kg]	S [mg/kg]	Na [mg/kg]	Al [mg/kg]	Fe [mg/kg]	Mn [mg/kg]	Zn [mg/kg]
X-P8	0-4	663	13492	634	3743	215	360	65708	39254	844	73
X-P8	4-50	569	13260	251	3406	168	328	66352	39536	835	67
X-P8	50-66	564	13950	159	4188	172	329	81700	46557	755	71
X-P8	66-	604	13496	187	4356	163	336	88109	53277	578	82

ECEC

Profile	Depth	ECEC [cmolc/kg]	BS [%]	H [µeq/g]	Na [µeq/g]	K [µeq/g]	Ca [µeq/g]	Mg [µeq/g]	Al [µeq/g]	Fe [µeq/g]	Mn [µeq/g]
X-P8	0-4	6.18	97.87	0.01	0.20	7.09	28.27	24.96	0.00	0.03	1.28
X-P8	4-50	3.40	72.39	0.00	0.23	4.82	7.66	11.88	8.00	0.04	1.35
X-P8	50-66	3.28	46.79	2.51	0.50	4.16	0.93	9.76	13.60	0.04	1.29
X-P8	66-	3.24	63.81	1.75	0.24	4.79	3.05	12.59	9.11	0.03	0.83

Table A 10: Data of profile 9 (Jiang Bian Zhan)

Soil physics

C and N

Profile	Depth	BD	BD air dry	BD oven dry	Sand	Silt	Clay	Texture	pH	SOC	TN	C/N	SOC	TN
		[g/cm³]	[g/cm³]	[g/cm³]	[%]	[%]	[%]		(H2O)	[g/kg]	[g/kg]		[t/ha]	[t/ha]
X-P9	0-7	1.16	1.06	1.05	27.62	33.94	38.43	CL	4.36	31.70	1.75	18.07	23.24	1.29
X-P9	7-24	1.41	1.25	1.24	31.48	35.22	33.17	CL	4.43	18.60	1.38	13.44	39.19	2.92
X-P9	24-74	1.41	1.19	1.18	21.74	38.66	39.57	CL	4.53	8.22	0.81	10.12	48.46	4.79
X-P9	74-	1.39	1.23	1.22	28.58	24.78	46.51	C	4.71	4.88	0.73	6.71	15.44	2.30

Nutrients

Profile	Depth	P	K	Ca	Mg	S	Na	Al	Fe	Mn	Zn
		[mg/kg]	[mg/kg]	[mg/kg]	[mg/kg]	[mg/kg]	[mg/kg]	[mg/kg]	[mg/kg]	[mg/kg]	[mg/kg]
X-P9	0-7	314	13086	233	859	163	776	48226	31362	129	21
X-P9	7-24	322	15112	213	943	141	867	55441	35901	119	25
X-P9	24-74	323	18423	272	1433	58	1045	87614	48374	114	29
X-P9	74-	292	19837	211	1534	54	1144	98728	54728	125	31

ECEC

Profile	Depth	ECEC	BS	H	Na	K	Ca	Mg	Al	Fe	Mn
		[cmolc/kg]	[%]	[µeq/g]	[µeq/g]	[µeq/g]	[µeq/g]	[µeq/g]	[µeq/g]	[µeq/g]	[µeq/g]
X-P9	0-7	5.57	10.77	3.19	0.66	2.59	1.66	1.08	45.08	0.31	1.09
X-P9	7-24	4.87	5.80	2.48	0.22	1.68	0.45	0.47	42.85	0.16	0.40
X-P9	24-74	5.41	5.22	2.62	1.00	1.54	0.29	0.00	48.51	0.12	0.04
X-P9	74-	6.18	2.16	3.13	0.36	0.97	0.00	0.00	57.29	0.05	0.01